527397/3.11.87/P/33.60

129.60

D1941759

DEVELOPMENTS IN FOOD FLAVOURS

An industry—university co-operation Symposium organised under the auspices of the National College of Food Technology (Department of Food Technology), University of Reading, 25–27 March, 1986

THE SYMPOSIUM COMMITTEE

GORDON G. BIRCH, B.Sc., Ph.D., D.Sc.(Lond.), F.R.S.C., C.Chem., M.R.S.H.
Reader, National College of Food Technology (Department of Food Technology), Food Studies Building, University of Reading, Whiteknights, Reading, Berks. RG6 2AP

M. G. LINDLEY, Ph.D.
Tate and Lyle PLC, Group Research and Development, Philip Lyle Memorial Research Laboratory, PO Box 68, Reading, Berks. RG6 2BX

K. J. PARKER, M.A., D.Phil. (Oxon), F.R.S.C., C.Chem.
'Murrens', Reading Road, Crowmarsh Gifford, Wallingford, Oxon. OX10 8EN

E. J. ROLFE, B.Sc., M.Chem.A., F.R.S.C., C.Chem., F.I.F.S.T.
Principal, National College of Food Technology (Department of Food Technology), Food Studies Building, University of Reading, Whiteknights, Reading, Berks. RG6 2AP

Mrs B. A. SHORE
c/o National College of Food Technology (Department of Food Technology), Food Studies Building, University of Reading, Whiteknights, Reading, Berks. RG6 2AP

Bibl.

DEVELOPMENTS IN FOOD FLAVOURS

Edited by

G. G. BIRCH and M. G. LINDLEY

FIRMENICH SA GENÈVE - Bibliothèque -

ELSEVIER APPLIED SCIENCE
LONDON and NEW YORK

TP 372.5.BIR 1986

ELSEVIER APPLIED SCIENCE PUBLISHERS LTD
Crown House, Linton Road, Barking, Essex IG11 8JU, England

Sole Distributor in the USA and Canada
ELSEVIER SCIENCE PUBLISHING CO., INC.
52 Vanderbilt Avenue, New York, NY 10017, USA

WITH 52 TABLES AND 98 ILLUSTRATIONS

© ELSEVIER APPLIED SCIENCE PUBLISHERS LTD 1986

British Library Cataloguing in Publication Data

Developments in food flavours.
1. Flavoring essences
I. Birch, G. G. II. Lindley, M. G.
664 TP418

Library of Congress Cataloging in Publication Data

Developments in food flavours.

'An industry–university co-operation symposium organised under the auspices of the National College of Food Technology, Department of Food Technology, University of Reading, 25–27 March, 1986'—P.
Includes bibliographies and index.
1. Flavor—Congresses. I. Birch, G. G. (Gordon Gerard) II. Lindley, M. G. III. National College of Food Technology, Dept. of Food Technology. IV. Title: Developments in food flavours.
TP372.5.D49 1986 664'.07 86-16765

ISBN 1-85166-037-2

3280/23. 11. 1987

The selection and presentation of material and the opinions expressed are the sole responsibility of the author(s) concerned.

Special regulations for readers in the USA
This publication has been registered with the Copyright Clearance Center Inc. (CCC), Salem, Massachusetts. Information can be obtained from the CCC about conditions under which photocopies of parts of this publication may be made in the USA. All other copyright questions, including photocopying outside of the USA, should be referred to the publisher.

All rights reserved. No part of this publication may be reproduced, stored in a retrieval system, or transmitted in any form or by any means, electronic, mechanical, photocopying, recording, or otherwise, without the prior written permission of the publisher.

Printed in Great Britain by Galliard (Printers) Ltd, Great Yarmouth

List of Contributors

J. ADDA

Laboratoire de Recherches sur les Aromes, Institut National de la Recherche Agronomique, 17 rue Rully, BV 1540, 21034 Dijon, France.

M. H. BOELENS

Destilaciones Bordas Chinchurreta SA, Apartado 11, 41008 Seville, Spain.

S. D. GANGOLLI

British Industrial Biological Research Association, Woodmansterne Road, Carshalton, Surrey SM5 4DS, UK.

L. J. VAN GEMERT

TNO-CIVO Food Analysis Institute, PO Box 360, 3700 AJ Zeist, The Netherlands.

J. KNIGHTS

Flavour Division, PFW Ltd, PO Box 18, Greenford, Middx UB6 7JH, UK.

G. MACLEOD

Department of Food and Nutritional Sciences, King's College, University of London, Campden Hill Road, London W8 7AH, UK.

D. A. MOYLER

Pauls Flavours & Fragrances Ltd, North Albert Road, Reigate, Surrey RH2 9ER, UK.

H. E. NURSTEN
Department of Food Science, Food Studies Building, University of Reading, Whiteknights, PO Box 226, Reading, Berks RG6 2AP, UK.

J. S. PAGINGTON
Cairn Foods, 60 High Street, Chesham, Bucks HP5 1EG, UK.

M. J. C. RHODES
Plant Cell Culture Group, AFRC Institute of Food Research, Colney Lane, Norwich NR4 7UA, UK.

P. SCHREIER
Lehrstuhl für Lebensmittelchemie, Universität Würzburg, Am Hubland, D-8700 Würzburg, Federal Republic of Germany.

Y.-H. SUGITA
Society for Research on Umami Taste, Kyobashi 1-5-8, Chuo-ku, Tokyo 104, Japan.

W. I. TAYLOR
International Flavors and Fragrances (EAME), Trafalgar House, Hammersmith, London W6 8DN, UK.

D. M. H. THOMSON
Department of Food Science, Food Studies Building, University of Reading, Whiteknights, PO Box 226, Reading, Berks RG6 2AP, UK.

F. B. WHITFIELD
CSIRO, Division of Food Research, PO Box 52, North Ryde, New South Wales 2113, Australia.

A. A. WILLIAMS
AFRC Institute of Food Research, Reading Laboratory, Shinfield, Reading, Berks RG2 9AT, UK.

Contents

PROCEEDINGS OF PREVIOUS SYMPOSIA

Glucose Syrups and Related Carbohydrates, edited by G. G. Birch, L. F. Green and C. B. Coulson, 1970.

Sweetness and Sweeteners, edited by G. G. Birch, L. F. Green and C. B. Coulson, 1971.

Health and Food, edited by G. G. Birch, L. F. Green and L. G. Plaskett, 1972.

Molecular Structure and Function of Food Carbohydrate, edited by G. G. Birch and L. F. Green, 1973.

Vitamin C, edited by G. G. Birch and K. J. Parker, 1974.

Food from Waste, edited by G. G. Birch, J. T. Worgan and K. J. Parker, 1975.

Intermediate Moisture Foods, edited by R. Davies, G. G. Birch and K. J. Parker, 1976.

Sensory Properties of Foods, edited by G. G. Birch, J. G. Brennan and K. J. Parker, 1977.

Sugar, Science and Technology, edited by G. G. Birch and K. J. Parker, 1978.

Food and Health: Science and Technology, edited by G. G. Birch and K. J. Parker, 1979.

Enzymes and Food Processing, edited by G. G. Birch, N. Blakebrough and K. J. Parker, 1980.

Nutritive Sweeteners, edited by G. G. Birch and K. J. Parker, 1981.

Dietary Fibre, edited by G. G. Birch and K. J. Parker, 1982.

Control of Food Quality and Food Analysis, edited by G. G. Birch and K. J. Parker, 1983.

Alcoholic Beverages, edited by G. G. Birch and M. G. Lindley, 1984.

Interactions of Food Components, edited by G. G. Birch and M. G. Lindley, 1985.

All published by Elsevier Applied Science Publishers.

1

The Meaning of Flavour

D. M. H. THOMSON

Department of Food Science, University of Reading, UK

ABSTRACT

Foods are complex mixtures of organic and inorganic compounds and, as such, are physico-chemical phenomena. During the process of consumption, various of the physico-chemical characteristics of foods selectively stimulate all our senses, with the consequence that we perceive a multitude of sensory (i.e. psychological) phenomena, collectively known as the sensory characteristics of foods, which it is commonplace to consider as comprising three mutually overlapping categories, termed flavour, texture and appearance. Flavour is defined by the British Standards Institution as 'The combination of taste and odour. It may be influenced by sensations of pain, heat and cold, and by tactile sensations'. The reason why the three categories should have evolved is unclear, as is their precise psychological identity. It is argued that conceptual dissociation of flavour from texture and appearance is arbitrary and probably a matter of mental abstraction. It may be that our thought processes are structured by familiarity with the words 'flavour', 'texture' and 'appearance', which artificially partition our conscious experiences. In summary, flavour is a poorly defined psychological phenomenon which is often confused with textural and visual characteristics. Sensory methods, rather than chemical methods, must be used to evaluate flavour, in the first place at least.

DEFINITION OF FLAVOUR

Flavour is defined by the British Standards Institution[1] as 'The combination of taste and odour. It may be influenced by sensations of pain, heat and cold, and by tactile sensations'. Other definitions of flavour[2–4] particularly emphasise the idea that flavour is a 'total' or 'unitary' experience that arises from stimulation of the sense of taste, the sense of smell and other senses.

These definitions imply that flavour is a phenomenon that occurs or arises when food and certain living organisms interact. Thus, in considering the meaning of flavour, it is pertinent to ask three questions. (1) How do foods interact with living organisms to cause the perception of flavour? (2) What is the nature of the phenomenon? (3) Why do we perceive flavour?

HOW DO FOODS INTERACT WITH LIVING ORGANISMS TO CAUSE THE PERCEPTION OF FLAVOUR?

Living organisms interact with objects that exist in the 'physical world', by way of their senses. Contrary to popular belief, there are more than five human senses. According to Marks,[5] the senses of taste, smell, sight, hearing and pain make five; if touch (including kinaesthesis), cold and warmth are considered collectively as a single somaesthetic sense, the total is six, but if these are considered separately the total is eight.

Foods are complex mixtures of organic and inorganic compounds. During the process of consumption, various physical and chemical entities within a food may selectively stimulate all our senses to some extent.

The first awareness of a food will usually come from smelling, seeing and/or hearing the food, probably at some distance, depending on the type of food and method of preparation. Next, the processes of cleaving a manageable 'bite size' portion from the food (as necessary) and then bringing the food up to the mouth for ingestion necessitate that the food should be handled in some way, either directly with the hands or indirectly with cutlery. Through this action, various of the somaesthetic 'senses', particularly touch and kinaesthesis, are stimulated, as are the senses of sight, smell and hearing. When the food is bitten and chewed, somaesthetic sensory receptors positioned on the lips, gums, tongue, throat and other inner surfaces of the mouth, including those sensitive to warmth and cold, are repeatedly stimulated. In addition, the chopping and grinding actions of the teeth disrupt the structure of the food matrix,

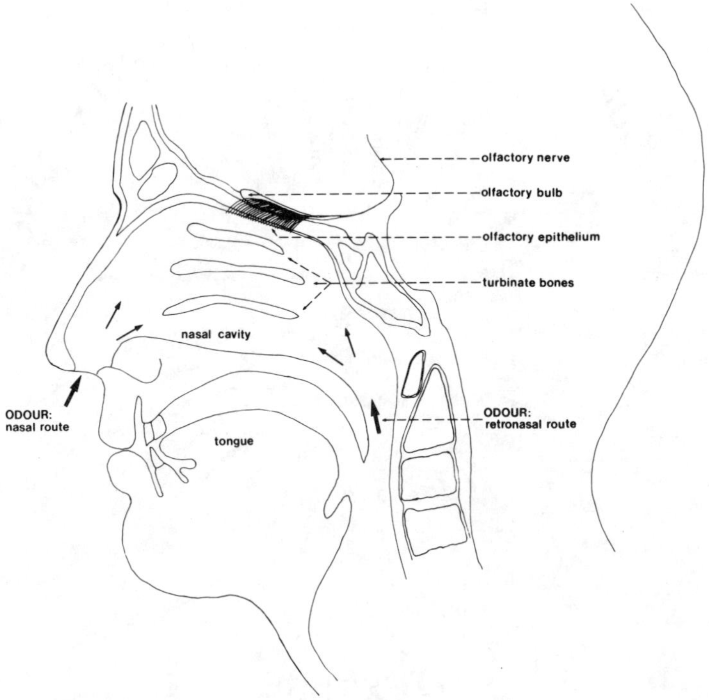

FIG. 1. Diagrammatic representation of a section through human nasal and buccal cavities.

rupture discrete cellular entities, and intimately mix the food material with saliva. During this process, soluble substances leach from the chewed material and some interact with the receptors of the sense of taste (gustatory cells). Volatile materials also escape from the food, and some of these are wafted into the nasal cavity on streams of exhaled and inhaled air, by way of the retronasal and nasal routes (see Fig. 1). A small proportion of these compounds will make contact with the receptor cells of the sense of smell in the olfactory epithelium (Fig. 1). Sound waves created when the food is bitten and chewed will be detected by the sense of hearing. Finally, somaesthetic receptors will also be stimulated when the food is swallowed.

The consequence of this onslaught of stimulation is that the individual will perceive a host of sensory phenomena, collectively known as the sensory characteristics of foods (Fig. 2). These sensory

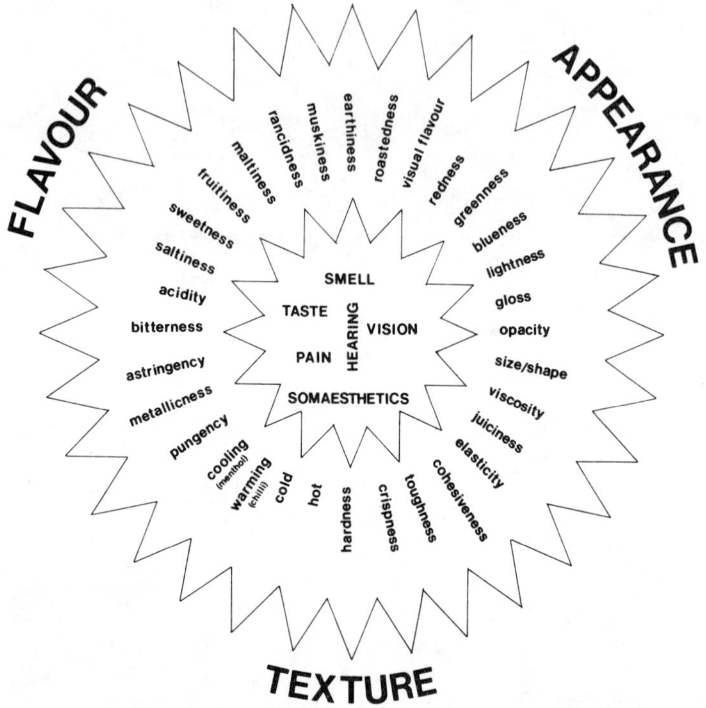

FIG. 2. Schematic representation of the sensory characteristics of foods.

phenomena are usually organised into three mutually overlapping categories, termed flavour, texture and appearance, the so-called principal sensory characteristics of foods.

Although the previous definitions of flavour implicate almost all, if not all, of the senses in flavour perception, the senses of taste (gustation) and smell (olfaction) play the leading part. For this reason consideration of the interaction between food components and sensory receptors will be restricted to taste and smell.

The Sense of Taste

The expression 'taste' is often used in common language to describe some, or possibly all, of the sensory characteristics perceived when a food is consumed.[6] However, when referring specifically to the sense of taste, perception of just four sensory characteristics is usually implied, namely saltiness, sweetness, acidity (sourness) and bitterness. In addition, Plattig[7] has described electric, alkaline and

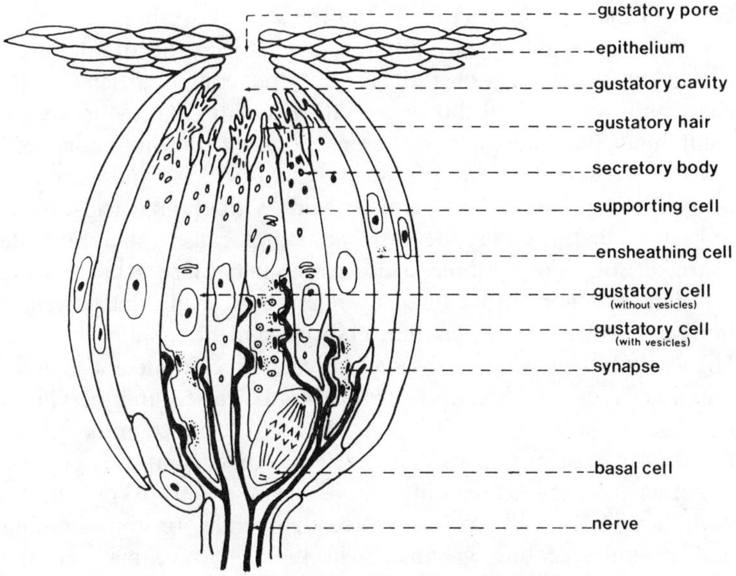

FIG. 3. Diagrammatic representation of a section through a human taste bud.

metallic 'subqualities' of taste. According to Plattig, electric taste relates to the perception of sourness, and alkaline or soapy tastes, when various parts of the tongue are stimulated by the positive and negative poles (respectively) of an electric DC source. However, since sourness itself is widely interpreted as being one of the four common tastes, the status of the subquality 'electric taste' is unclear. Metallic taste is apparently caused by some metal salts and formed metals, whilst the alkaline/soapy taste is elicited by potassium carbonate, as well as cathodal DC stimulation.

The peripheral receptors of the sense of taste are the gustatory cells, which are located within discrete organelles, familiarly known as taste buds (Fig. 3). Estimates of the number of taste buds on the human tongue vary from 2000 to 9000, with each housing 15–18 receptor cells. Taste buds are, in turn, located within specialised structures called papillae, which occur mainly on the tip, sides and rear of the upper surface of the tongue. Papillae effectively increase the surface area of the tongue and are responsible for its characteristic rough appearance.

To cause a taste sensation, stimulus molecules must somehow gain access to the gustatory cells. Since the contents of the gustatory cavity of the taste buds (Fig. 3) and the saliva covering the tongue are aqueous, solubility in water is a characteristic of all taste-causing molecules. However, the features that confer particular taste characteristics on a substance are rather less obvious. Most of the compounds that cause acid taste sensations are proton donors, and mineral acids at the same pH, such as hydrochloric and sulphuric acids, are perceived as being equally acid. With organic acids, other molecular features may be influential and cause different taste characteristics. For example, picric acid usually causes a bitter taste. The salt taste is mainly caused by relatively low molecular weight inorganic salts, such as sodium chloride and potassium chloride. Higher molecular weight salts such as caesium chloride and potassium iodide cause bitterness,[8] whilst lead acetate and beryllium chloride cause sweetness. A feature believed to be common to the diverse range of molecules that cause sweetness comprises proton-donating and proton-accepting groups spaced exactly 0·3 nm apart.[9] Kubota and Kubo[10] have suggested that similar proton-donating and proton-accepting groups, spaced at 0·15 nm, may be the molecular moiety that confers bitterness.

The mechanisms through which taste stimuli interact with gustatory receptor cells, and ultimately cause transmission of an electrical signal to the brain, are far from clear. It has been hypothesised by Beidler[11] that the sensorially 'active' feature of a stimulus molecule somehow binds with a corresponding receptor site on the exposed surface of a gustatory cell. This alone, or in conjunction with other such events, may cause an alteration in the potential difference that normally exists across the gustatory cell membrane at rest, and ultimately induces an action potential in an adjacent neurone. In order to explain perception of the four common taste qualities, it was for a long time held that taste buds, nerve fibres, papillae, and even areas of the brain, were taste quality specific. However, these ideas have been largely discredited, since it has been widely demonstrated[12] that gustatory cells within a single taste bud can apparently elicit a variety of taste qualities, when suitably stimulated. Other theories, such as the 'across fibre pattern hypothesis', postulate that the various taste qualities are coded as a 'pattern' of electrical impulses, transmitted simultaneously by several nerve fibres.[13] Presumably, the brain recognises and processes these 'patterns', so that one or other of the taste qualities is ultimately perceived.

Sensitivity to taste-causing stimuli varies within and across indi-

TABLE 1
Typical Recognition Thresholds (Minimum Concentration at which a Stimulus is Correctly Identified)

Compound	*Perceived phenomenon*	*Stimulus concn (% wt/vol H_2O)*
Sodium chloride	Saltiness	0·1
Sucrose	Sweetness	0·5
Caffeine	Bitter	0·015
Tartaric acid	Acid	0·002

viduals, and 100-fold variation within a population is common. Typical recognition threshold values for sodium chloride (saltiness), sucrose (sweetness), caffeine (bitterness) and tartaric acid (acidity) are shown in Table 1.

The Sense of Smell

The human sense of smell is qualitiatively more diverse than the sense of taste, and is characterised by an almost infinite number of odour qualities. The sense of smell is also much more sensitive than taste, in that the former can detect certain volatile compounds at concentrations as low as 10^{-18} molar. Characteristic acute sensitivity to a particular compound usually diminishes very quickly after the onset of stimulation, often to the point of complete insensitivity. However, once the stimulus has been withdrawn, recovery is usually rapid. Although individuals vary enormously in their olfactory sensitivity, this variation can only be rather loosely associated with a few broad factors. For example, a general diminution of olfactory sensitivity may occur with advancing age. This may be attributable to the loss of primary olfactory nerve fibres, at the rate of 1% per year after birth.

According to Harper[14] there are more than 17 000 chemical compounds that are potential odorants. Many of these are included in a comprehensive listing of odour stimuli threshold values, compiled by Stahl.[15]

The peripheral receptors of the olfactory sense are special neurones which are localised in a relatively small area of the mucous-covered inner surface of the nasal cavity, called the olfactory epithelium (Figs 1 and 4). Receptor cells are characterised by a swollen tip (olfactory vesicle), from which protrude some 6–12 hair-like

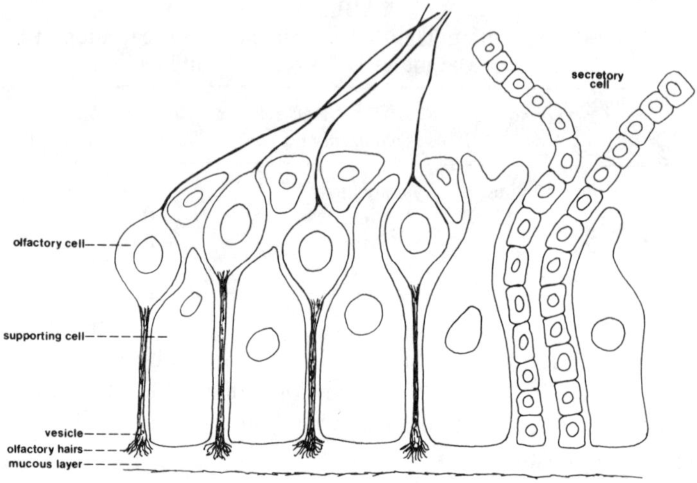

FIG. 4. Diagrammatic representation of a section through part of the human olfactory epithelium.

features.[16] The function of these so-called olfactory hairs is not known for certain but, if nothing else, they effectively increase the surface area of the receptor cells and hence the entire olfactory epithelium. Estimates as to the number of such receptor cells in man vary from 5×10^7 [17] to 3×10^6.[18]

Volatile odorant compounds normally access the nasal cavities, and hence the sensitive olfactory epithelium, by way of the outer nose (nasal route; Fig. 1) and the throat (retronasal route; Fig. 1), in streams of inhaled and exhaled air. However, only a very small proportion of odorous substances that enter the nasal cavity in this way actually make contact with the sensitive receptor cells. Odorants can also reach the olfactory epithelium by way of the bloodstream, by diffusing from capillaries in the lungs and nasal cavities.[19] Blood-borne odorants may have an important role in the regulation of food intake.

The most widely accepted explanation of the assumed interaction between stimulus molecule and olfactory receptor cell is the 'membrane puncture theory' of Davies.[20] It is hypothesised that stimulus molecules are adsorbed onto the surface of the receptor sites and then desorbed, leaving a momentary breach in the receptor cell membrane, through which ionic interchange may occur. This would effectively change the potential difference across the receptor cell membrane, which could in turn induce an action potential in an adjacent neurone. The number of molecules required to desorb from a receptor site simultaneously, in order to breach the membrane, would vary according to the compound and may explain the observed differences in threshold values.

As with the gustatory sense, a seemingly diverse range of compounds can cause qualitatively similar olfactory sensations. If it is assumed that the physico-chemical properties of any molecular species, under specific physical conditions, constitute the only source of information carried by the molecules,[21] then it can be reasoned that molecular species which elicit qualitatively similar olfactory sensations must have some physico-chemical property in common. There have been many attempts to correlate physico-chemical properties with odour quality (e.g. Schiffman *et al.*[22]), but none has been conclusive.

From the 'Physical World' to the 'Psychological World'

To conclude, food interacts with human beings to cause flavour, by impinging on almost all our senses, but particularly taste and smell. Certain physico-chemical properties of the foods cause the peripheral receptors of the various senses to generate action potentials in adjacent neurones, and these electrical signals are ultimately transmitted to various parts of the brain. It is in the brain that the 'information' originally held by the food is somehow transformed from the 'physical world' to the 'psychological world'.

FLAVOUR: WHAT IS THE NATURE OF THE PHENOMENON?

When living organisms consume or otherwise interact with foods, a mass of sensory information is transmitted to the brain, and some of this sensory input will probably result in the perception of flavour. In order to understand how the perception of flavour arises, and to appreciate the nature of this sensory phenomenon, it is essential to consider how the brain processes all of this sensory input.

A Model for the Interpretation of Sensory Input

A four-stage interactive model is proposed which describes the major cognitive processes involved in the interpretation of sensory input (Fig. 5).

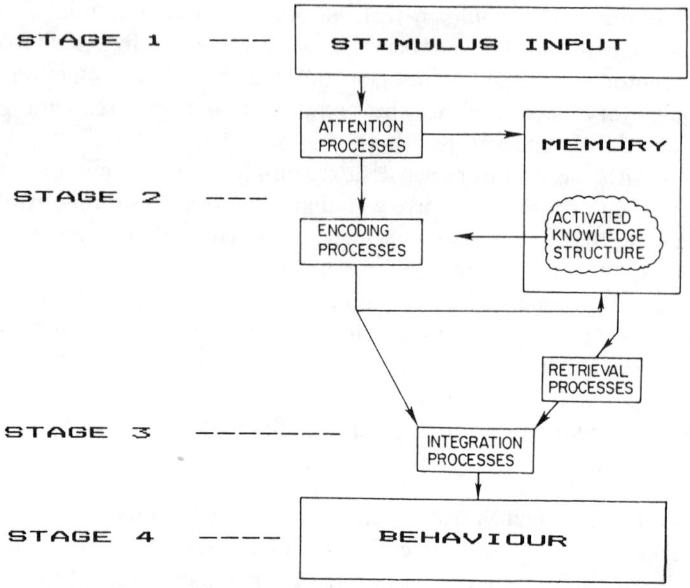

FIG. 5. Interactive model of the major cognitive processes in the interpretation of sensory input (modified from Olson[23]).

Stage 1

The trigger that initiates this sensory process exists in the 'physical world'. For the sake of simplicity the physical trigger can be arbitrarily divided into three categories: (1) the food itself; (2) the absolute physical amount of effort and/or money that the consumer actually will have to expend/pay to obtain the food, and transform it into the desired state for consumption; (3) physical information about the environment and circumstances in which the food and the consumer interact. When interaction between food and consumer occurs, the physical information is first translated into neural impulses (still in the physical world) and then transmitted to the brain. It is particularly important to appreciate that the sensory input will also comprise information not directly related to the food as such, but still important in terms of the food consumption process.

Stage 2

Here, the perceiving organism attends to the sensory input (attention processes; Fig. 5), which presumably requires the expenditure of some amount of cognitive processing effort. This is followed by an initial identification stage (encoding processes; Fig. 5), where the input information is interpreted and identified in the light of existing knowledge that has been specifically activated in the memory structure, by the sensory input, on that specific occasion (so-called 'activated knowledge structure'; see Fig. 5). An individual's momentary interpretation of the input will clearly depend on previously acquired meanings that already exist in the activated knowledge structure.[24]

The result at this stage is a psychological interpretation of the physical world. Three categories of perception (i.e. psychological characteristics) are envisaged, which correspond to the three categories of physical information previously described: food, cost and environment/circumstances. Food factors comprise the information received and recalled by the brain regarding the nature of the food, such as the sensory characteristics of the food (flavour, texture and appearance; see Fig. 2) or its package, compositional information, and any other corresponding information received and recalled as a consequence of input arising from advertising, packaging and display. Cost factors relate to the individual's evaluation of the cost, in terms of money and/or physical effort, of acquiring the food and transforming it into the desired state for consumption. Environmental/circumstantial factors relate to the individual's interpretation of the place, the time of day and the company in which the food is purchased, prepared and consumed.

Some information-processing theorists[25] believe that knowledge exists in the memory in the form of 'memory schemata', where each schema consists of a limited amount of acquired knowledge organised into a framework. During the course of a lifetime, an individual will acquire a vast library of such memory schemata, which is presumably subject to constant revision.[26] It is envisaged that when the memory is activated by sensory input the appropriate memory schemata are called up and used to identify and interpret the incoming information.

Since all individuals differ, to some extent, in the information they hold in their memories, and hence in the structure of their various memory schemata, it is highly likely that individuals will also differ in their interpretation of the sensory input. Harper[27] has emphasised the importance of such individual differences, in food acceptance research, market research and sensory evaluation.

Stage 3

The next stage proposed in the model is the integration of information referred to above with further information retrieved from the memory (retrieval and integration processes; Fig. 5). This results in the individual making value judgements regarding the food/food situation (i.e. hedonic evaluation of sensory characteristics, evaluation of cost, evaluation of consumption environment/circumstances, evaluation of nutritional efficacy, evaluation of image, etc.). Clearly, these value judgements will depend on that individual's evaluative and judgemental criteria, as evidenced by the wide differences that exist amongst individuals in their choice and selection of foods. Hence, some people will consider the food to have a pleasant flavour, whilst others consider the flavour to be less pleasant; some will consider the food to be 'worth the money' whilst others would disagree.

Stage 4

Finally, the individual's interpretation, evaluation and judgement regarding the food/food situation will ultimately determine their overt behaviour (i.e. behavioural response, behavioural intention or attitude) in response to the original physical input.

Stimulus–Organism–Response

Stages 1 and 4 of the foregoing model relate to phenomena that are measurable or observable in the physical world. Stages 2 and 3 relate to occurrences in the private psychological world of each individual; as such, they are fundamentally unobservable. Thus Stage 1 describes the *stimulus*/stimulatory processes, Stages 2 and 3 describe the internal psychological events of the *organism*, and Stage 4 represents the organism's ultimate *response* to the stimulus, in the light of newly acquired and existing knowledge. This whole process is sometimes referred to as the S–O–R (stimulus–organism–response) paradigm of psychophysical judgement.[28–30]

The Flavour of Foods

Flavour, and indeed all the sensory characteristics of foods, constitute only a fraction of the perceptual phenomena recognised by an individual when food is consumed (refer to Stage 2 of Fig. 5). It is essential to recognise that an individual's response to the sensory characteristics that are bracketed together as flavour will undoubtedly be influenced by the other ('non-flavour') sensory characteristics of the food, and all the other 'non-food' sensory phenomena that are simultaneously perceived, when a food is consumed. Thus there are likely to be severe contextual shortcomings when flavour is singled out and considered in isolation from these other sensory phenomena.

The reason why the sensory characteristics of foods are bracketed together under the headings of flavour, texture and appearance (Fig. 2) is unclear, as is their precise psychological status. However, it may at least be deduced that flavour, texture and appearance are not 'equivalent' psychological phenomena. This is evident from the observation that the overall intensity of flavour can be evaluated in terms of 'How strong is the flavour?'. In contrast, texture is usually quantified by, for example, 'How tough . . .?', 'How juicy . . .?', 'How hard . . .?', etc., with 'How light . . .?', 'How intensely red . . .?', 'How glossy . . .?', etc., for appearance (see Fig. 2). Intuitively this suggests that flavour may be perceived as a single integrated phenomenon (sensory complex), whereas the various sensory phenomena implicated in texture and appearance are perceived independently.

Apart from the foregoing, there is an insufficiency of empirical evidence to confirm or deny that flavour is perceived as a sensory complex. In order to test this point, several criteria could be applied. For example, Garner and co-workers[31] have suggested that complex sensory phenomena are processed in a 'wholistic scan' where the global characteristics of the perceived phenomenon are emphasised over the underlying constituent dimensions. Also, the Gestalt criterion, that the whole (flavour) is more than the sum of its parts (constituent sensory characteristics), could perhaps be tested.[17]

Attempts to find discrete identities for flavour, texture and appearance are confounded by the fact that some sensory phenomena seemingly contribute to more than one of these principal sensory characteristics. For example, colour obviously contributes to the appearance of a food, but it may also contribute to flavour. It can be readily demonstrated that variation in the colour of a sweet-tasting solution, whilst all other factors remain constant, will normally change the perceived sweetness intensity.[32] Also, lemon-flavoured jelly which is of the expected colour will normally be described as 'lemon' or 'citrus'; colour the same jelly scarlet with a flavourless food dye, and many people will adamantly describe the flavour as 'raspberry' or 'strawberry'. This probably occurs by virtue of the fact that the existing information held in the memory strongly associates red-coloured jelly with strawberry or raspberry flavour (i.e. particular memory schemata; see Fig. 5), and the contradictory information derived from the red 'lemon' jelly is initially overridden or ignored. The observation that a stimulus can sometimes cause the perception of an unexpected, or indirectly related, sensory phenomenon is comprehensively documented in the psychological literature.

Interaction between appearance and flavour is sometimes called 'visual flavour'. Similarly, sensory phenomena which are normally associated with texture can influence the perception of flavour and vice versa.[33] Interaction between visual and textural phenomena can also occur.

On the basis of these observed overlaps, conceptual dissociation of flavour, from what might normally be considered as textural or visual sensory phenomena, may be arbitrary and possibly a matter of mental abstraction. Indeed, it may be that our thought processes are merely structured by familiarity with the words 'flavour', 'texture' and 'appearance', which artificially partition our conscious experiences. It is particularly noteworthy that, in the Dutch language, there is no direct translation of 'flavour', and taste and smell are typically considered as discrete sensory phenomena.

Thus, from existing evidence, precise psychological interpretation of flavour is probably impossible. However, the following points are worth noting:

(1) Flavour exists exclusively in the psychological world.
(2) Psychological phenomena such as flavour cannot be observed or measured directly in the physical world.
(3) Flavour arises when a stimulus (e.g. food or beverage) interacts with a living organism.
(4) Flavour comprises a number of sensory characteristics, largely derived by stimulation of the olfactory and gustatory senses.
(5) It is uncertain if flavour is a complex sensory phenomenon, or merely a loose aggregation of individual sensory characteristics, combined on an arbitrary basis.

Measurement and Evaluation of Flavour

Since flavour is a psychological phenomenon and not a physico-chemical phenomenon, it can only be evaluated or measured by sensory methods. In using sensory methods, a human subject is typically asked to translate perceived psychological phenomena into words and/or numbers, distances, or other analogues of perceptual magnitude. According to the foregoing S–O–R paradigm, these words and numbers should be interpreted as 'responses', because they are behavioural manifestations of the interaction between stimulus and organism. In practice, considerable caution is essential when interpreting such words and numbers, because they may be imprecise or inappropriate representations of phenomena that exist exclusively in the psychological world. To this end, there are a number of excellent texts that explain the principles and practices of sensory evaluation in detail.[2,4,34]

Direct evaluation of flavour, by physico-chemical methods, is fundamentally impossible. At best, such methods can only qualify and quantify certain physico-chemical properties of the stimulus, which may then be related to the various responses obtained by sensory methods. However, since flavour seems to be a rather vague combination of psychological phenomena, and foods are extremely complex mixtures of organic and inorganic compounds, the chances of qualitatively and quantitatively identifying all the various components of the stimuli that cause a particular flavour are extremely remote. This onerous task can be simplified if food aroma rather than flavour is considered, but even then the relationships may be complex and obscure.[35]

WHY DO WE PERCEIVE FLAVOUR?

It might be logically assumed that humans have evolved, or at least utilised, the sensory mechanisms through which they perceive flavour for a specific purpose, possibly linked with survival.

Instinctive Behaviour

The 'survival hypothesis' is supported, in part, by the observation that humans, especially the very young, often show reflexive behaviour to compounds that cause sweetness and bitterness. There is a considerable amount of evidence which suggests that this reflexive behaviour may be innate. For example, De Snoo[36] has described how the injection of saccharin into the amniotic fluid surrounding a 3–4 month old human foetus caused the rate of sucking-like motions in the foetus to increase markedly. Immediately after birth, the human infant will typically display a distinct affinity to sweetness,[37] and water is commonly sweetened to induce drinking in new-born infants. In marked contrast, bitter and sour tasting solutions are typically rejected by young humans. The fact that certain highly toxic substances and poisonous berries are intensely bitter may be significant in this respect. Instinctive behaviour has also been demonstrated towards certain odours. For example, Steiner[38] has described how new-born babies display appropriate facial expressions to food-related odours, even prior to any feeding experience.

Learned Associations

As infants grow and mature, instinctive behaviour is often modified in the light of experience. For example, moderate levels of bitterness are widely appreciated by adults, in coffee, dark chocolate, bitter lemon, beer, gin and tonic, and other foods and beverages. Behaviour modification of this type occurs through learning. For example, repeated sampling of bitter-tasting foods, with no apparent adverse post-ingestional consequence, may eventually cause consumers to modify their interpretation of bitterness. However, it may be relevant that most 'acceptable' bitter foods (see above) contain other powerful metabolic stimuli (e.g. alcohol, caffeine, sugar, etc.), and it may be that the 'benefit' derived from these substances outweighs natural aversion to bitterness.

During the course of a lifetime, but especially during the early formative years, associations between taste, odour or flavour and post-ingestional consequence are learned and augmented, and these learned associations greatly influence food choice. Learned associations have been convincingly demonstrated by Barker,[39] in a taste aversion conditioning experiment using laboratory rats. Several independent groups of rats were allowed to taste and ingest small volumes of a dilute saccharin solution. Lithium chloride was then injected into the rats, 0·5, 2·0 or 21·0 hours later, with adverse consequence for the rats. On recovery, the rats were given the opportunity to express their preference or aversion for saccharin, by choosing to drink either the dilute saccharin solution or water. The rats experiencing the adverse effects of lithium chloride, either 0·5 or 2·0 hours after saccharin ingestion, showed avoidance of saccharin. Presumably the rats associated the sensory characteristics of saccharin with the adverse post-ingestion experience. It would have been very interesting to find out how long this learned association was sustained without further reinforcement.

There are compelling logical reasons for presuming that the association described between sensory characteristics (particularly flavour) and adverse visceral consequence extends almost universally to the post-ingestional consequences of foods. For example, Blundell[40] proposed that, once the decision to commence eating has been made, for whatever reason, choice of available food is partly directed by the anticipation of the pleasure to be derived from the flavour and other sensory characteristics of the food. Similarly, the rate of food ingestion, and the amount of each food actually consumed, which Blundell describes as manifestations of appetite, are seemingly governed by change in the perceived pleasantness of the sensory characteristics of the food during consumption. Change in the pleasantness of flavour (or any other perceived phenomenon) with change in the internal milieu is termed alliesthesia.

Everyday experience and formal experimental evidence[41] indicate that alliesthesia, and hence appetite and satiety, are sensory specific.[42] For example, most people will recall occasions where they have been completely satiated by the meat course of a meal, but able to consume a sizeable portion of a sweet dessert. A possible explanation for so-called sensory specific satiety is that it encourages variety in food intake, and hence the acquisition of a greater variety of nutrients.

Implicit in the foregoing is the assumption that associations between the sensory characteristics of foods and consequent changes in the internal milieu are anticipated in the light of previous experiences and do not arise from immediate physiological effects. For example, it is assumed that the diminution of flavour pleasantness, which causes curtailment of appetite, signals physiological sufficiency. However, since the metabolic consequences of food are unlikely to be apparent until some time after the meal has ended, it must be assumed that the change in flavour pleasantness, and the consequent loss of appetite, occurs in anticipation of the metabolic effect and hence is learned. In short, it would seem that the flavour of a food is analogous to a label, which allows the organism to predict the metabolic consequences of ingesting that particular food,

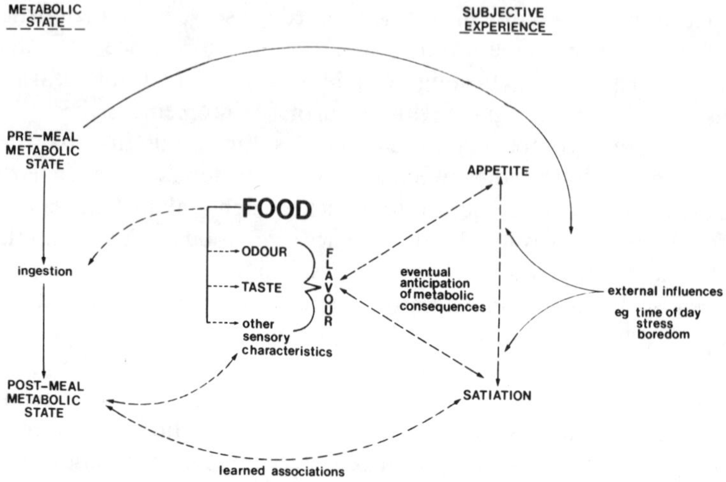

FIG. 6. Schematic representation of the relationships formed between the metabolic consequences of eating, the sensory characteristics of foods, and the subjective experiences that accompany eating.

and to take anticipatory action to adjust the amount consumed.[43] Possible relationships formed between the metabolic consequences of eating, sensory characteristics of food, and the subjective experiences that accompany eating are set out schematically in Fig. 6.

Flavour Pleasantness and the Control of Body Weight

Cabanac[44] has described a possible relationship between the perceived pleasantness of the taste and smell of a food and the momentary body weight of an individual. Central to this theory is the idea that people are 'programmed' by external and internal factors to reach a specific weight, bulk, or level of fatness. If the body weight is lower (or higher) than the programmed weight, the flavour of the food is perceived as being more (or less) pleasant than if the momentary weight of the individual is at the so-called 'set-point'.

Other theories relating flavour pleasantness and body weight control have evolved around the experimental observation that 'obese' subjects generally consumed more food than 'normal' weight subjects, when the food was perceived as being highly palatable.[45] When the expressed degree of liking for the food was relatively low, there was no significant difference in the amount of food eaten by the two groups. This suggests that obese people may be influenced to a greater extent than their normal weight peers, by so-called 'external cues' such as the flavour of a food, and possibly to a lesser extent by 'internal cues' such as visceral satiety signals. However, any conclusions drawn from these observations must also take account of the fact that people who are highly responsive to external cues are not always obese.

Flavour and Nutrition

Flavour plays a very large part in our enjoyment of food. Since flavour is, superficially, more directly associated with enjoyment than nourishment, it has often been given low priority by nutritionists, if not completely disregarded. This is a serious oversight, since even the desperately hungry will reject food that has an undesirable or unfamiliar flavour. The appalling record of diet adherence in those whose very survival may depend on a careful selection of nutrients (e.g. diabetics) is not difficult to understand, when the foods they are expected to eat may have undesirable, monotonous or unfamiliar flavours. People cannot realistically be expected to override a lifetime of learned flavour associations overnight, irrespective of their current state of health. Consequently, the likely success of a therapeutic diet depends not only on the nutritional efficacy of the food but also on its general acceptability and flavour.

Flavour: Challenge for the Future

Since flavour plays a major role in the day-to-day and momentary regulation of food intake, manipulation or modification of the human diet for whatever reason (e.g. implementation of dietary guidelines, or for reasons of commercial expedience) is unlikely to succeed without taking flavour into account. Indeed, flavour could be a particularly useful medium through which to achieve recommended changes in eating patterns. Thus the challenge for food science and technology is to produce foods that are profitable, nutritionally efficacious and emotionally satisfying. Flavour may be the key to success in this respect.

REFERENCES

1. British Standards Institution (1975). *British Standard Glossary of Terms Relating to the Sensory Analysis of Food,* British Standards Institution, London.
2. Amerine, M. A., Pangborn, R. M. and Roessler, E. B. (1965). *Principles of Sensory Evaluation of Food,* Academic Press, New York and London.
3. Rothe, M. (1972). *Nährung,* **16**, 473–81.
4. Jellinek, Gisella (1985). *Sensory Evaluation of Food: Theory and Practice,* Ellis Horwood, Chichester, UK.
5. Marks, L. E. (1978). *The Unity of the Senses,* Academic Press, New York.
6. Rubin, E. (1936). *Brit. J. Psychol.,* **27**, 74–85.
7. Plattig, K. H. (1984). In: *Sensory Analysis of Foods,* Ed. Piggott, J. R., Elsevier Applied Science, London, 1–22.
8. Teranishi, R., Hornstein, I., Issenberg, P. and Wick, Emily L. (1971). *Flavor Research: Principles and Techniques,* Marcel Dekker, New York.
9. Shallenberger, R. S. and and Acree, T. E. (1967). *Nature,* **216**, 480–2.
10. Kubota, T. and Kubo, I. (1969). *Nature,* **223**, 97–9.
11. Beidler, L. M. (1954). *J. Gen. Physiol.,* **38**, 133–9.
12. Plattig, K. H. and Innitzer, J. (1976). *Pfluegers Arch.,* **361**, 115–20.
13. Pfaffman, C., Frank, M., Bartoshuk, L. M. and Snell, T. C. (1976). In: *Progress in Psychobiology and Physiological Psychology,* Eds Sprague, J. M. and Epstein, A. N., Academic Press, New York, 1–27.
14. Harper, R. (1982). *Food Flavourings, Ingredients, Packaging, Processing,* **4**, 13–21.
15. Stahl, W. H. (1973). *Compilation of Odor and Taste Threshold Values Data,* American Society for Testing and Materials, Baltimore.
16. Moulton, D. G. and Beidler, L. M. (1967). *Physiol. Rev.,* **47**, 1–52.
17. Harper, R. (1972). *Human Senses in Action,* Churchill Livingstone, London.
18. Moran, D. T., Rowley, J. C., Jafek, B. W. and Lovell, M. A. (1982). *J. Neurocytol.,* **11**, 721–46.
19. Maruniak, J. A., Silver, W. L. and Moulton, D. G. (1983). *Brain Res.,* **265**, 312–16.
20. Davies, J. T. (1965). *J. Theor. Biol.,* **8**, 1–7.
21. Beets, M. G. J. (1978). In: *Handbook of Perception. VIA. Tasting and Smelling,* Eds Carterette, E. G. and Friedman, M. P., Academic Press, New York, 245–55.
22. Schiffman, S. S., Robinson, D. E. and Erickson, R. P. (1977). *Chem. Senses Flav.,* **2**, 375–90.
23. Olson, J. C. (1981). In: *Criteria of Food Acceptance,* Eds Solms, J. and Hall, R. L., Forster Verlag, Zürich, 69–81.
24. Norman, D. A., Rumelhart, D. E. and LNR Research Group (1975). *Explorations in Cognition,* Freeman, San Francisco.

25. BOBROW, D. G. and NORMAN, D. A. (1975). In: *Representation and Understanding: Studies in Cognitive Science,* Eds Bobrow, D. G. and Collins, A. M., Academic Press, New York.
26. MARKUS, H. (1977). *J. Pers. Soc. Psychol.*, **35**, 67–78.
27. HARPER, R. (1981). In: *Criteria of Food Acceptance,* Eds Solms, J. and Hall, R. L., Forster Verlag, Zürich, 220–37.
28. TORGERSON, W. S. (1961). *Acta Psychol.,* **19**, 201–5.
29. ATTNEAVE, F. (1962). In: *Psychology: A Study of Science,* Ed. Koch, S., McGraw-Hill, New York.
30. TREISMAN, M. (1964). *Quart. J. Exp. Psychol.,* **16**, 11–22.
31. GARNER, W. R. and FELFODY, G. L. (1970). *Cognitive Psychol.*, **1**, 225–34.
32. MAGA, J. (1974). *Chem. Senses Flav.,* **1**, 115–21.
33. THOMSON, D. M. H. (1981). *An Investigation of Non-verbal Sensory Techniques for Evaluating Meat Flavour,* University of Bristol.
34. STONE, H. and SIDEL, J. L. (1985). *Sensory Evaluation Practices,* Academic Press, New York.
35. NURSTEN, H. E. (1979). In: *Progress in Flavour Research,* Eds Land, D. G. and Nursten, H. E., Applied Science, London, 337–55.
36. DE SNOO, K. (1937). *Monats. Gerbutschilfe,* **105**, 88–97 (in German; cited by Beidler[37]).
37. BEIDLER, L. M. (1982). In: *The Psychobiology of Human Food Selection,* Ed. Barker, L. M., Ellis Horwood, Chichester, UK, 3–15.
38. STEINER, J. E. (1977). In: *Taste and Development: The Genesis of Sweet Preference,* Ed. Weiffenbach, M., US Dept. of Health, Education and Welfare (Publ. No. NIH 77-1068), Bethesda, Maryland.
39. BARKER, L. M. (1982). In: *The Psychobiology of Human Food Selection,* Ed. Barker, L. M., Ellis Horwood, Chichester, UK, 85–99.
40. BLUNDELL, J. E. (1979). In: *Nutrition and Lifestyles,* Ed. Turner, M. R., Applied Science, London, 21–42.
41. CABANAC, M. (1971). *Science,* **173**, 1103–7.
42. ROLLS, B. J., ROLLS, E. T. and ROWE, E. A. (1982). In: *The Psychobiology of Human Food Selection,* Ed. Barker, L. M., Ellis Horwood, Chichester, UK, 101–22.
43. BOOTH, D. A., LEE, M. and MCALEAVEY, C. (1976). *Brit. J. Psychol.,* **67**, 137–47.
44. CABANAC, M. (1979). In: *Preference Behaviour and Chemoreception,* Ed. Kroeze, J. H. A., Information Retrieval, London, 275–89.
45. NISBETT, R. E. (1968). *J. Pers. Soc. Psychol.,* **10**, 107–16.

2

Physicochemical Parameters Related to Organoleptic Properties of Flavour Components

MANS H. BOELENS
Destilaciones Bordas Chinchurreta SA, Seville, Spain

and

LEO J. VAN GEMERT
TNO-CIVO Food Analysis Institute, Zeist, The Netherlands

ABSTRACT

Straight chain, saturated and unsaturated, aldehydes are important constituents of citrus, fruit, dairy, vegetable, meat and other flavours. For instance, alkanals and 2-alkenals (C_6 to C_{14}) have been detected and quantified in Bitter Orange Oil. The organoleptic qualities and threshold values of these compounds are quite different; moreover, the qualities change a lot on dilution. An attempt has been made to correlate the physicochemical parameters with the organoleptic properties of alkanals, alkenals and alkadienals from C_6 to C_{14}. The studied physicochemical parameters are molecular weight, vapour pressure, melting point, molecular refraction, partition coefficient and dipole moment; the structural features have also been studied. Organoleptic qualities (of concentrated and diluted stimulants) and threshold values have been investigated. Structure–activity relationships have been found for the flavour components, alkanals, alkenals and alkadienals, with isolated and conjugated double bonds. The organoleptic qualities include penetrating-pungent, green-herbal, rose-fatty, citrusy and others. An explanation is advanced of the effects of *length of carbon chain, place of double bond(s), isolated and conjugated, and Z and E geometrical isomers, on organoleptic quality and threshold values.*

INTRODUCTION

During human odour and taste perception one may assume that a chemical substance (or mixture) interacts with a biological system affording a response. Moreover, one can assume that the response is correlated to the structural features and the physicochemical parameters of the chemical(s) and the properties of the biological system. With respect to the organoleptic properties of the stimulus, the responses commonly express something about detection (yes/no), recognition (qualitatively), intensity (weak/strong) and preference (like/dislike). The structural features and physicochemical parameters of chemical compounds can be defined by two factors, the stereocity and the electronicity of the stimulating molecules. The stereocity concerns the complete profile of the molecule including molecular size, shape and chirality, subdivided into molecular weight, molecular volume, occurrence of enantiomers, diastereoisomers, epimers and geometrical isomers [*cis*/*trans* = *Z*(usammen)/*E*(ntgegen) isomerism]. The electronicity concerns the complete electronic charge distribution within the molecule, provided by molecular orbital calculations, and the particular polarity (polarisability) expressed for instance by dipole moments or dielectric constants. The electronicity is mainly determined by the functional group(s), in relation to the rest of the molecule.

The physicochemical parameters of stimulating molecules concern the volatility, the transport and the interaction possibilities of the chemical. Physicochemical parameters in use are molecular weight, vapour pressure, partition coefficient, molecular refractivity, dipole moment, retention time in chromatography, etc. One should notice that these parameters often are interrelated.

The following sections will be devoted to a review of the literature regarding the organoleptic properties of aldehydes, some aspects of the compounds under study, some aspects of threshold values, an attempt to correlate the physicochemical parameters of aldehydes with their threshold values, study of the relationships between structural features and organoleptic qualities of aldehydes, and final remarks and conclusions.

LITERATURE REVIEW

Because this study is mainly based on literature data, it seems worthwhile to give a review of some important publications regarding the organoleptic properties of aldehydes.

Straight chain saturated and unsaturated fatty aldehydes have been found in almost every food flavour.[1] The olfactive odour-qualities are mentioned in several books and publications.[2–7] Bedoukian mentioned, with respect to the olfactive properties of alkanals, that it is remarkable that octanal, decanal and dodecanal have odours akin to each other, the differences being largely the result of changes in molecular weight. The aldehydes having odd numbers of carbon atoms in their molecules are quite different in character. The organoleptic flavour-qualities in mineral oil have been determined by Meijboom and Jongenotter[8] and those in beer by Meilgaard.[9]

To evaluate the importance of aldehydes many scientists have measured odour and flavour thresholds. The flavour threshold values of hexanal, octanal, decanal and *trans*-2-nonenal have been determined in water, groundnut oil and paraffin oil by Lea and Swoboda.[10] Day *et al.*[11] investigated the flavour threshold values of a series of alkanals (C_5 to C_{10}), *trans*-2-alkenals and of *trans,trans*-2,4-heptadienal in milk. The flavour thresholds of alkanals (C_5 to C_8) in oil and water have been published by Siek *et al.*[12]

For a study of the relationship between molecular structure and flavour perceptibility of aliphatic aldehydes Meijboom[13] determined the threshold values of odour and taste of 10 alkanals (C_3 to C_{12}), 8 *trans*-2-alkenals (C_5 to C_{12}), *cis*- and *trans*-3-hexenal, and 11 alkadienals in solutions of paraffin oil. A definite relation between the threshold value and the chain length of the compound could not be concluded. In all cases the threshold values for flavour were lower than those for odour. It was noted that, for aldehydes also, the gradually more accepted rule applies that the taste assessment in fact takes place in the odour receptors of the nose, and consequently can be qualified as odour. The *trans*-2-alkenals with an odd number of carbon atoms had lower thresholds than the corresponding aldehydes with even numbers. With *trans,trans*-2,4-alkadienals just the opposite held true. Meijboom found this alternating effect in various series of odour and flavour experiments, but an explanation for this phenomenon has not been found. He stated that it is generally known that various homologous series display alternating effects in the case of certain physical properties, for instance the melting points and heats of crystallisation. Such alternating effects usually refer to a physical property of the crystalline state of the substance in question. Therefore it is not impossible that, during odour and taste perception, molecular orientation plays a part. Moreover he found, with mixtures of various aldehydes, that, for certain ratios, an antagonistic effect between odour and flavour occurred.

Later, Meijboom and Jongenotter[14] studied the flavour thresholds of unbranched isomeric nonenals in paraffin oil. They found that the odour threshold values of the isomeric *trans*-nonenals decrease when the double bond moves from the carbonyl function, with a pronounced minimum for *trans*-6-nonenal. With flavour threshold values, an alternating effect was observed, with the lowest value for *trans*-6-nonenal again. With the isomeric *cis*-nonenals, both odour and flavour thresholds alternated, with a minimum for *cis*-6-nonenal.

More recently Meijboom and Jongenotter[8] investigated the flavour perceptibility of straight chain, unsaturated aldehydes as a function of double-bond position and geometry. The threshold values for odour and flavour of 34 straight chain mono- and of 5 di-unsaturated aldehydes were determined in solutions of paraffin oil. The threshold values were low and displayed interesting odour and flavour aspects. Special attention was given to the influence of chain length, double-bond position, number of double bonds and *cis/trans* geometry. They found alternating odour and flavour threshold values for the *trans*-$(n-x)$-heptenals and *cis*-$(n-x)$-nonenals. The homologous series of *trans*-2,*trans*-$(n-3)$-alkadienals had one minimum for the odour and flavour threshold value, viz. for *trans*-2,*trans*-6-nonadienal. Earlier investigations,[15] in which the thin-layer chromatographic behaviour of these aldehydes was studied as 2,4-dinitrophenylhydrazone derivatives on silver-nitrate-impregnated chromatoplates, showed similar results. No answer was found to which way these thin-layer chromatographic phenomena can relate to the odour and flavour aspects of unsaturated aldehydes in the free state. They mentioned that basic organoleptic investigations of these compounds, using animal odour and taste receptors, might better elucidate the mechanism of smelling and tasting.

Recently Dixon and Hammond[16] studied the flavour intensity of some carbonyl compounds important in oxidised fats. They evaluated the flavour intensities of several alkanals, *trans*-2-alkenals, *trans*-2,*trans*-4-alkadienals and methyl ketones in 1% mineral oil in water emulsions against a series of standard emulsions of 2-heptanone, giving an equal flavour intensity. Threshold values calculated from these linear relations were comparable to those reported in the literature. Their data for the threshold values of *trans*-2-alkenals and *trans*-2,*trans*-4-alkadienals showed some alternation but did not support Meijboom's[13] generalisations.

Guadagni *et al.*[17] studied odour thresholds of aldehydes associated with food flavours. A simple technique for measurement of olfactory thresholds of odorous materials was described. They found that the thresholds of a homologous series of saturated aliphatic aldehydes decreased progressively with increase in chain length from C_5 to C_{10}. Similar results were obtained with a series of unsaturated aldehydes. They stated that Henry's law predicts that partial pressure is directly proportional to concentration in very dilute solutions. This was tested and found to be true for several aldehydes in water solution over the range 10 ppb to 10 ppm. They found an essentially linear relationship between the logarithm of the threshold concentration in molecules per ml and chain length from about C_5 to C_{10}. It seemed apparent that compounds with higher molecular weight (from C_5 to C_{10}) required fewer molecules to stimulate the olfactory mechanism. It also seemed apparent that conversion of the weight concentrations in water solution to concentrations in the vapour phase did not significantly change the order of results.

Later, Guadagni *et al.*[18] determined the odour thresholds of the major components found in steam-volatile potato chip oil, in water and in vegetable oil. These components e.g. included some alkanals, *trans*-2-alkenals and *trans*-2,*trans*-4-decadienal.

Buttery *et al.*[19] studied the volatilities of aldehydes, ketones and esters in dilute water solution. In a later publication[20] they studied the volatilities of organic compounds in foods.

With respect to the aforementioned investigations,[17–20] Teranishi *et al.*[21] studied odour quality and chemical structure in fruit and vegetable flavours, which included alkanals (C_2 to C_{11}), *trans*-2-alkenals (C_3 to C_{10}), *cis*-3-hexenal, *trans*-4-hexenal, 2-*trans*,4-*trans*-nonadienal and 2-*trans*,6-*cis*-nonadienal. The threshold values of these aldehydes in air were calculated from the threshold values in water and the air/water partition coefficient. They arrived at the following general patterns.

(a) Most alkanals and *trans*-2-alkenals require roughly the same order of concentration of molecules in air for significant detection.
(b) With alkenals a double bond not in conjugation with the carbonyl generally results in lower thresholds than one in conjugation.
(c) *trans*-2,*trans*-4-Alkadienals generally have lower thresholds than the corresponding *trans*-2-alkenals.
(d) Alkanals and *trans*-2-alkenals generally have lower thresholds than the corresponding 2-alkanones and 2-alkenones.

For each of these patterns, however, there are exceptions. They mentioned that a double bond conjugated with the carbonyl group increases the polarity of the molecule. This increases the water-solubility of the compound as well as other properties. A double bond not conjugated with the carbonyl would be expected to have very little effect on the general physical properties. Going from a *trans*-2-alkenal to a *trans*-2,*trans*-4-alkadienal increases the length of conjugation and hence increases the polarity of the molecule. *trans*-2,*trans*-4-Alkadienals are therefore more soluble in water than *trans*-2-alkenals, and their air/water partition coefficients are lower.

More recently Amoore and Buttery[22] made a thorough study of partition coefficients and olfactometry. They described a gas-flow method for obtaining equilibrium concentration data to calculate the coefficient of certain odorants. They also made a comparison between calculated and observed air/water partition coefficients of aliphatic aldehydes (C_3 to C_9).

Hall and Andersson[23] determined odour thresholds and odour intensity functions of volatile fat oxidation products by dynamic olfactometry. The relationship between perceived odour intensities and concentrations of single compounds was determined, e.g. for alkanals (C_3 to C_9), *trans*-2-alkenals (C_4 to C_{10}) and *trans*-2,*trans*-4-alkadienals (C_7 to C_9). Subsequently they transformed odour thresholds, obtained by making dilution series in water, into concentrations in the gas phase by multiplying concentrations in water by experimentally determined gas/water partition coefficients.

Very recently Christoph and Drawert[24] investigated structure–activity relationships by studying olfactory thresholds of compounds, including alkanals C_6 to C_{11}, determined by a gas chromatographic sniffing technique. For these aldehydes they concluded that the threshold values tend to decrease with increasing number of carbon atoms. They recognised a linear decrease, but threshold minima for hexanal, octanal and undecanal. They referred to the minimum threshold for decanal, calculated by Teranishi *et al.*[21] They mentioned that in spite of the different medium, water, corresponding investigations by Rothe *et al.*[25] also show minima for octanal and for decanal, while Schnabel[26] found the same minima as in their study.

With respect to threshold values in different media, several publications refer to partition coefficients. Backmann[27] in 1917 published a note about the odour potential of compounds and their solubility in water and oil. He stated that compounds with a perceptible odour have to be sufficiently soluble in oil and also sufficiently soluble in water. Moreover, he derived a hypothesis which states that, for the odour potential of homologous series and isomers, the relation between the solubility in water and in lipids is important, which he demonstrated with several examples.

For the calculation of odour thresholds in air and in solutions the partition coefficient air/solvent has to be known. Amoore and Buttery[22] mentioned that Raoult's law, which attempts to predict the value of the partition coefficient from the mole fractions of the substance (solute) and of the solvent, may be completely misleading. Henry's law, however, which relates the solubility of a gas to its partial pressure, is usually reliable. They mentioned that, though many solutions of odorants are non-ideal in the classical sense (i.e. Raoult's law is not obeyed), yet the less ambitious Henry's law and the general distribution law are followed accurately enough for most olfactometric applications.

Gramshaw and Williams,[28] in their physicochemical studies on flavour-active compounds, wrote that, if it is necessary to know the minimum concentration of an odorant in air which is required to stimulate response, this is readily available, at least in principle, as the product of the threshold, measured as an aqueous solution, and the air/water partition coefficient, $K_{A/W}$. It is possible, however, that potent odorants may exhibit surface activity and thus become concentrated into the interface between atmosphere and aqueous solution. If this were to occur, $K_{A/W}$ would increase as the concentration decreased, and more odorant molecules woiuld be released into the gas phase than predicted from the accepted value $K_{A/W}$ (which would have been measured using a solution concentration many orders of magnitude greater than the threshold concentration). They developed a method to demonstrate that 2-ethyl-3-methoxypyrazine showed an increase in the value of $K_{A/W}$ with dilution, whilst 2-isopropyl-3-methoxypyrazine did not possess this behaviour.

MATERIALS AND PHYSICOCHEMICAL DATA

This study concerns the structure–activity relationships of about 70 straight chain alkanals, alkenals and alkadienals with 6 to 14 carbon atoms. These unsaturated linear fatty aldehydes (C_6 to C_{14}) can be subdivided into 9 normal alkanals, 126 *cis*- and *trans*-alkenals and 669 *cis*- and *trans*-alkadienals. From these 804 aldehydes only about 10% are known in the literature.

Linear saturated and unsaturated fatty aldehydes have been found in a lot of food flavours, such as milk and other dairy products, animal fats and vegetable oils, fruit and citrus flavours. Very recently we have detected and quantified 18 alkanals, *trans*-2-alkenals and *trans*-2,4-*cis*/*trans*-alkadienals from C_6 to C_{14} in the peel oil of bitter orange (*Citrus aurantium* L. subspecies *amara* L.). The total concentration of these aldehydes in the peel oil was 1·4%. Some of the aldehydes in 0·1% solution evoked the characteristic orange peel organoleptic quality. The relative retention times ($C_8 = 100$; $C_{14} = 200$) of these aldehydes were determined on a gas-chromatographic glass capillary column with an apolar silicon oil (SE-54) phase (see Table 1).

Aldehydes are rather reactive and can undergo oxidation, hydration, acetalisation, trimerisation and aldol condensation. The most important decomposition reactions are probably oxidation (with air) and trimerisation which occurs with traces of acids. Because trace impurities can disturb the organoleptic qualities of the aldehydes, the products (under study) should be purified carefully.

All physicochemical parameters in Tables 1 and 2 are from

TABLE 1
Physicochemical Parameters of Aldehydes, Regarding the Transport of Molecules

Aldehyde	*Mol. weight*	*Boiling point (°C)*	*BP/MW*	*Vapour pressure (mm Hg)*	*Water solubil. (25°C) (g/l)*	$K_{A/W}$ (1×0^{-3})	*log K*	*log P octanol/water*	*Relative GC retention times*	*Threshold[a] value* T_{air} $(10^{-6}\ g/l)$
C_6	100·2	128·2	1·279	9	5	9·3	−2·03	1·60	67·6	0·053
C_7	114·2	152·6	1·336	3	1·5	13·8	−1·86	2·12	82·2	0·037
C_8	128·2	172	1·342	1	0·5	20·5	−1·69	2·64	100	0·011
C_9	142·2	191	1·343	0·4	0·1	30	−1·52	3·16	118·2	0·008
C_{10}	156·3	209	1·334	0·1	0·025	45	−1·35	3·68	136·4	0·001
C_{11}	170·3	223	1·310	0·04	(0·006)	66	−1·18	4·20	153·5	0·004
C_{12}	184·3	237	1·286	0·02	(0·002)	102	−1·01	4·72	170·0	0·002
$C_6 2E$	98·1	151	1·539	(4)	(50)	1·59	−2·8	1·3	—	0·034
$C_7 2E$	112·2	166	1·480	(1·5)	(15)	2·51	−2·6	1·8	—	0·034
$C_8 2E$	126·2	(190)	(1·51)	(0·5)	(5)	4·0	−2·4	2·3	110·6	0·009
$C_9 2E$	140·2	(210)	(1·50)	(0·15)	(2)	6·35	−2·2	2·8	129·1	0·000 5
$C_{10} 2E$	154·2	(230)	(1·49)	(0·03)	(0·5)	10·0	−2·0	3·3	146·1	0·002 7

Values in parentheses are estimated.
[a] Geometric mean of all determined literature data.[29,30]

TABLE 2
Physicochemical Parameters of Aldehydes, Regarding the Stereocity and Electronicity of Molecules

Aldehyde	*Mol. weight*	*Molar vol.*	*Mol. vol. C number*	*Melting point (°C)*	*Molar refractivity*	$\frac{n^2-1}{n^2+2}$	*Dipole moment* μ	*Threshold value* T_{air} (10^{-6} *g/l*)
C_6	100·2	123·1	20·5	−56	30·1	0·245	(2·7)	0·053
C_7	114·2	138·9	19·8	−45	34·5	0·249	(2·5)	0·037
C_8	128·2	156·0	19·5	(−27)	39·7	0·254	(2·7)	0·011
C_9	142·2	172·2	19·1	(−24)	44·2	0·257	(2·5)	0·008
C_{10}	156·3	187·6	18·8	−5	48·5	0·258	(2·7)	0·001
C_{11}	170·3	205·9	18·7	−4	53·3	0·259	(2·5)	0·004
C_{12}	184·3	219·9	18·3	+11	57·4	0·261	(2·7)	0·002
$C_6 2E$	98·1	115·6	19·3	—	30·9	0·268	(3·7)	0·034
$C_7 2E$	112·2	129·7	18·5	—	33·6	0·259	(3·5)	0·034
$C_8 2E$	126·2	147·6	18·4	—	37·6	0·254	(3·7)	0·009
$C_9 2E$	140·2	161·7	18·0	—	43·1	0·267	(3·5)	0·000 5
$C_{10} 2E$	154·2	182·5	18·3	—	48·7	0·270	(3·7)	0·002 7

Values in parentheses are estimated.

the literature[3–5] (*Handbook of Chemistry and Physics*; *Beilstein's Handbuch*). Values between brackets have been estimated by extrapolation or interpolation from known data, e.g. boiling points at different pressures, known melting points and dipole moments for homologues (C_1 to C_4).

For determination of the organoleptic qualities n-alkanals (from our own production) were purified to a content of over 99%. Alkenals and alkadienals were obtained from Bedoukian Research Inc. An expert panel of five perfumers and flavourists judged the quality of the products in pure form and in a 0·1% solution in glycerine triacetate (triacetin).

THRESHOLD VALUES

The published threshold values for a particular compound often differ by an order of magnitude or even more.[29,30] This wide range of values for one and the same compound has several origins. Threshold values are determined by psychophysical methods, which can be separated into psychological and physical terms. The psychological part for instance concerns the selection of the subjects, their representativity (age, sex, occupation), conditioning (experience, motivation, advance knowledge), reproducibility, variability and reliability. The psychological part of the method can be improved sometimes, to a certain extent, but this seldom happens. The physical part of the method concerns, e.g. purity of the compound, concentration adjustment, presentation of the samples, interpretation of results, etc. This part of the method often can be improved in exact ways, so that systematic and accidental errors can be decreased or eliminated.

One would like to have a (physical) check on whether a threshold value determined for a certain chemical is more or less accurate. This physical check can be obtained for a homologous series of compounds by determination of the threshold values in different media and using the same systematic method. The quotient of the threshold values of one compound in two media equals the partition coefficient for this compound over the two media.

Odour threshold values experimentally determined in water can be converted to those in air and vice versa. If a compound is present in water and in equilibrium with the air phase, the concentrations in both phases are related to each other as follows:

$$K = \frac{\text{concentration in air}}{\text{concentration in water}} \tag{1}$$

in which K is the air/water partition coefficient. This relation can only be used to convert threshold values if the K-value is known or can be calculated from the vapour pressure, the water solubility and the activity coefficient, if these are reported in the literature. Otherwise the K-value has to be determined:

$$K = \frac{\text{odour threshold value in air}}{\text{odour threshold value in water}} \tag{2}$$

Data thus obtained by conversion will show less accuracy than those obtained by direct experimental measurements. The experimental K-value, according to eqn (2), can be verified with that of eqn (1):

$$K_{(1)} = K_{(2)} = \frac{\text{vapour pressure at } T\,°\text{C}}{\text{water solubility at } T\,°\text{C}} \quad \text{(in g/l)} \tag{3}$$

We derived the following formulas, which may have practical value for homologous series:

$$K = \frac{\text{threshold value in medium A}}{\text{threshold value in medium B}} = \frac{\text{concentration in A}}{\text{concentration in B}} \tag{4}$$

TABLE 3
Verification of Odour Threshold Values in Air and Water of n-Alkanals

Alkanal	*Threshold value in air*[a] *(10^{-6} g/l)*	*10^{-3} K*	*log K*	*Threshold value in water (10^{-6} g/l)*	
				calc.	*found*[17]
C_6	0·053	9·3	−2·03	6	4·5
C_7	0·037	13·8	−1·86	3	3·0
C_8	0·011	20·5	−1·69	0·5	0·7
C_9	0·0075	30	−1·52	0·3	1·0
C_{10}	0·0011	45	−1·35	0·2	0·1
C_{11}	0·0042	66	−1·18	0·1	0·5

[a] Geometric mean of all determined literature data.[29,30]

TABLE 4
Verification of Odour Threshold Values in Air and Water of *trans*-2-Alkenals

trans-2-Alkenal	*Threshold value in air*[a] *(10^{-6} g/l)*	*$10^{-3}K$*	*log K*	*Threshold value in water (10^{-6} g/l)*	
				calc.	*found*[17,21]
C_6	0·217	1·59	−2·8	137	17
C_7	0·106	2·51	−2·6	42	13
C_8	0·018	4·0	−2·4	4·5	3·0
C_9	0·013	6·35	−2·2	2	0·08
C_{10}	0·0037	10·0	−2·0	0·37	0·3

[a] Geometric mean of all determined literature data.[29,30]

$K = \text{Quotient}_x^a$, a = activity constant, x = functionality.

$$K = Q_x^a = f(Q_{(\mathrm{H(CH_2)}_n)}) + f(Q_{(\mathrm{functionality})}) \qquad (5)$$

$$\log K = a \log Q_x = n \log(\mathrm{CH_2}) + \log(\mathrm{functionality}) \qquad (6)$$

In homologous series like straight chain alkanals or 2-alkenals the functionality is constant. In those series there exists a linear relation between the logarithm of the partition coefficient and the number of carbon atoms, and as a consequence the logarithm of the increment per CH_2 group is constant.

Odour and flavour threshold values for aldehydes have been determined in various media, e.g air, water, vegetable oil, mineral oil, water/mineral oil emulsion and milk. In Tables 3–9 the threshold

TABLE 5
Verification of Odour Threshold Values in Vegetable Oil and Water of Alkanals

Alkanal	*Threshold value in*		*K*	*log K*
	veg. oil[17,18] *(10^{-6} g/l)*	*water*		
C_6	120	4·5	26·6	1·42
C_7	250	3	83	1·91
C_8	(430)[a]	0·7	(300)[a]	(2·48)[a]
C_9	1000	1·0	1000	3·0

[a] Calculated from interpolated log *K*.

TABLE 6
Verification of Odour Threshold Values in Vegetable Oil and Water of *trans*-2-Alkenals

trans-2-Alkenal	*Threshold value in*		*K*	*log K*
	veg. oil	*water*		
	(10^{-6} g/l)			
C_7	1500	13	115	2·06
C_8	500	3	167	2·22
	(1370)[a]		(457)[a]	(2·66)[a]
C_9	150	0·08	1875	3·27
C_{10}	2100	0·3	7000	3·85

[a] Calculated from interpolated log *K*.

values of alkanals and *trans*-2-alkenals in different media are verified with *K*-values and the log *K* values. It should be noted that log *K* values could deviate from linearity for lower (less than C_5) and higher (over C_{12}) homologues in a series, because of the dominant influence of the functional group or the alkyl group, respectively.

The aforementioned calculations are in agreement with the calculations of fragment and π-values from the partition coefficient (*P*), for the octanol-1/water system, which is used in pharmacochemistry to calculate the hydrophobic factor during drug action.

TABLE 7
Verification of Flavour Threshold Values in Milk and Water of Alkanals

Alkanal	*Threshold values (10^{-3} g/l) in*			*K*[a]	*log K*
	milk[11]	*water*[16]			
		found	*calc.*		
C_5	0·13	0·07–0·08	0·076	1·7	0·23
C_6	0·05	0·016–0·3	0·016	3·1	0·49
C_7	0·12	0·031	0·021	5·6	0·75
C_8	0·46	0·005	0·045	10·2	1·01
C_9	0·22	—	0·012	18·6	1·27
C_{10}	0·24	0·007	0·007	33·9	1·53

[a] Partition coefficient *K* is theoretical, because milk and water are miscible.

TABLE 8
Verification of Flavour Threshold Values in Mineral Oil and Water of Alkanals

Alkanal	*Threshold values (10^{-3} g/l) in*			*K*	*log K*
	mineral oil[13,16]		*water*		
	found	*calc.*	*calc.*[a]		
C_5	0·15–0·30	0·228	0·076	3·0	0·48
C_6	0·15–0·60	0·094	0·016	5·9	0·77
C_7	0·04–0·75	0·242	0·021	11·5	1·06
C_8	0·07–0·90	1·008	0·045	22·4	1·35
C_9	0·3	0·524	0·012	43·7	1·64
C_{10}	0·60–1·00	0·596	0·007	85·1	1·93

[a] See Table 7 for calculation.

TABLE 9
Calculation of Flavour Threshold Values in Water of *trans*-2-Alkenals

trans-2-Alkenal	*Threshold values (10^{-3} g/l) in*			*K*	*log K*
	min. oil[13]	*water*			
	found	*found*	*calc.*		
C_6	2·5	—	0·45	5·5	0·74
C_7	0·63	—	0·08	8	0·90
C_8	1·0	—	0·09	11·5	1·06
C_9	0·1	0·006	—	16·6	1·22
C_{10}	5·5	—	0·23	24	1·38

PHYSICOCHEMICAL PARAMETERS AND THRESHOLD VALUES

During odour and taste perception, in principle four steps can be recognised as primary processes of olfaction:

(1) Transport of the odorant.
(2) Interaction of the odorant with active biological sites (parts of receptors).
(3) Fit (or measurement) of the odorant at active sites.
(4) Pulse generation at the receptor.

With respect to these four steps the physicochemical properties can be divided into parameters regarding:

(a) transport: volatility (e.g. vapour pressures and partition coefficients);
(b) interaction: polarity (e.g. dipole moments and dielectric constants);
(c) fit: shape (molecular weight/volume, chirality);
(d) pulse generation: all properties of stimulus and acceptor (active site at receptor).

It should be noted that several physicochemical parameters are interrelated, e.g. molecular weight, molar volume, boiling point, molar refractivity and partition coefficient (*P*) over octanol-1/water. Therefore it does not make much sense to correlate each of these parameters with the threshold values.

Molecular Weight and Molar Volume

The molecular weight is most often used as a physicochemical parameter for structure–activity relationships studied in chemoreception. In homologous series, the molecular weight is replaced by the number of carbon atoms. We have plotted the threshold values in air of alkanals against their molecular weights in Fig. 1.

Guadagni *et al.*[17] concluded that there was an essentially linear relationship between the logarithm of the threshold concentration in molecules per ml of water and chain length from about C_5 to C_{10}. They mentioned that it seemed apparent that compounds with higher molecular weight, from C_5 to C_{10}, require fewer molecules to stimulate the olfactory mechanism. However, one year later Meijboom[13] stated that a definite relation between the flavour threshold value in paraffin oil and the chain length of aldehydes (alkanals, *trans*-2-alkenals and *trans*-2,*trans*-4-alkadienals) cannot be concluded; a conclusion might therefore be that, at least in the interval C_9 to C_{12}, the number of molecules required to stimulate the olfactory receptors decreases with increasing molecular weight.

In spite of the fact that molecular weights can be calculated easily and accurately, and therefore may be useful physicochemical parameters, the size of a molecule is better represented by its molar volume. The molar volume is the quotient of the molecular weight and the specific gravity of the compound. A disadvantage of the determination of this parameter is that the density of the compound

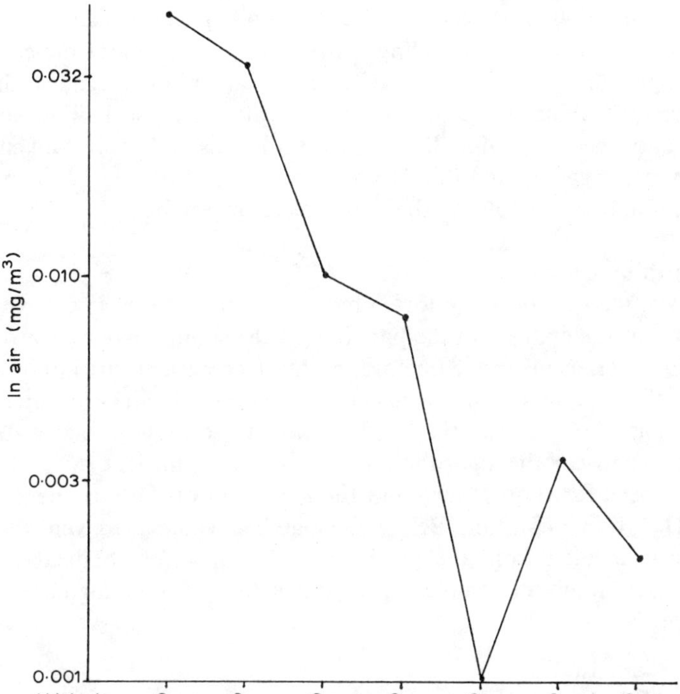

Aldehyde	C_6	C_7	C_8	C_9	C_{10}	C_{11}	C_{12}
MW	100	114	128	142	156	170	184
BP	128	153	172	191	209	223	237°C
MR	30·1	34·5	39·7	44·2	48·5	53·3	57·4

FIG. 1. Correlation of threshold values with molecular weights, boiling points and molar refractivities of n-alkanals.

must be known or determined and that it is dependent on temperature. Therefore the accuracy of the determination of the molar volume is less than that of the molecular weight. In Table 2 the molar volumes of alkanals and *trans*-2-alkenals and their volumes per C unit are shown. The correlation of molar volumes with threshold values can be compared with that of the molecular weights.

Volatility

The volatility of odorants at a given temperature is important for transport and as a consequence for the concentration of the compound at the olfactory epithelium. The determination of the vapour pressure of compounds at lower temperature (ambient or body) is difficult, especially for those of high boiling products.

As the volatility decreases and the boiling points increase in a homologous series, the boiling points are a function of the molecular weight. The boiling points of alkanals are plotted against their threshold values in air in Fig. 1. We calculated the boiling point fraction per mole unit (BP/MW) for alkanals and *trans*-2-alkenals which are shown in Table 1. The correlation of boiling points with threshold values follows that for molecular weights.

Partition Coefficients

For the transport of odorant molecules through the water/protein layer at the epithelium the partition of the stimuli may play a role. Moreover, as mentioned before, partition coefficients are important for the comparison and verification of threshold values in different media. We have seen that, in homologous series, a linear relation exists between the logarithm of the partition coefficient (K) and the number of carbon atoms, and the logarithm of the increment per CH_2 group is constant. Below, the log(CH_2) values are given, which are calculated from the partition coefficients (K) obtained from experimentally determined threshold values of homologous series (see Tables 3–9).

Alkanals:

Odour: $\log(CH_2)_{\text{air/water}} = 0{\cdot}17$

$\log(CH_2)_{\text{vegetable oil/water}} = 0{\cdot}50$

Flavour: $\log(CH_2)_{\text{mineral oil/water}} = 0{\cdot}29$

$\log(CH_2)_{\text{milk/water}} = 0{\cdot}26$

(this value is theoretical; milk and water are miscible)

trans-2-Alkenals:

Odour: $\log(CH_2)_{\text{air/water}} = 0{\cdot}20$

$\log(CH_2)_{\text{vegetable oil/water}} = 0{\cdot}60$

Flavour: $\log(CH_2)_{\text{mineral oil/water}} = 0{\cdot}16$

The partition coefficient (P) of compounds over octanol-1 and water is used in pharmacochemistry; the fragment value for alkanals and alkenals per CH_2 group: $\log P_{(CH_2)} = 0{\cdot}52$. The $\log P$ values are depicted in Table 1; they follow a similar correlation with threshold values as the molecular weights.

Melting Points

It is difficult to understand that a correlation between melting points and organoleptic properties of aldehydes could exist.[13] In Table 10 the differences in melting points and the differences in

TABLE 10
Correlation of Threshold Values with Melting Points

Going from aldehyde	*Diff. in melting point (°C)*	*Diff. in threshold value (air)*	*Diff. in threshold value (water)*
		(10^{-6} g/l)	
C_6 to C_7	11	0·016	1·5
C_7 to C_8	18	0·026	2·3
C_8 to C_9	3	0·003	0·3
C_9 to C_{10}	19	0·007	0·9
C_{10} to C_{11}	1	0·003	0·4
C_{11} to C_{12}	15	0·002	0·3
Average	11	0·009 5	0·95

threshold values (in air) for alkanals are shown. From this table it may be concluded that some correlation between melting points and threshold values exists.

Retention Times in Gas Chromatography

Retention times in chromatography are related to volatility, adsorption and desorption constants and partition coefficients. The retention times of 18 aldehydes in gas chromatography were determined on a silicon oil (SE-54) liquid phase. The relative times ($C_8 = 100$; $C_{14} = 200$) are shown in Table 1. The correlation of the retention times with threshold values can be compared with that of the molecular weights (Fig. 1). The correlation of retention times with threshold values within isomers with the same carbon number (C_{10}) is not quite clear (see Table 11).

Molar Refractivity

The molar refractivity $[(n^2 - 1)/(n^2 + 2)](M/d)$ is used in structure–activity studies in pharmacochemistry. The molar refractivity consists of two factors. The factor M/d means that molar refractivity depends completely on the molar volume. The term $(n^2 - 1)/(n^2 + 2)$ affords an indication of the mobility of the electrons in the molecule, e.g. the ease with which they are polarisable. The correlation of the molar refractivity with threshold values can be compared with that of the molecular weights (see Fig. 1).

Dipole Moments

The dipole moments for odd and even numbered C-chain aldehydes can differ by about 0·2 Debye units. Ethylenic double

TABLE 11
Correlation of Threshold Values (TV) with GC Retention Times

Aldehyde	TV_{odour} *in water* (*g/10 ml*)	TV_{odour} *in air* (*g/10 ml*)	*GC retention time*	TV_{odour} *in min. oil* (*ppm*)	$TV_{flavour}$ *in min. oil* (*ppm*)
Decanal	0·1	5×10^{-3}	136·4	6·7	1·0
trans-2-Decenal	0·3	$2·7 \times 10^{-3}$	146·1	33·8	5·5
trans-2,*trans*-4-Decadienal	0·07	$0·18 \times 10^{-3}$	151·6	2·15	0·28
trans-2,*cis*-4-Decadienal	—	—	155·4	(0·7)	(0·03)

TABLE 12
Correlation of Threshold Values with Dipole Moments

Aldehyde	*Dipole moment* μ	*Threshold value in air* (mg/m^3)
Methanal	2·33	0·184
Ethanal	2·69	0·080
Propanal	2·52	0·048
2-Propenal	3·12	0·3
Butanal	2·72	0·039
trans-2-Butenal	3·67	0·5

bonds conjugated with a carbonyl group increase the dipole moment by approximately 1 Debye unit; non-conjugated double bonds decrease it by about 0·5 Debye units. See Tables 2 and 12.

According to Kumler and Fohlen,[31] biological activity is not correlated with dipole moment. However, from Table 12 and within isomers with the same carbon number (C_9), some correlation between the dipole moments and threshold values can be observed (see Table 13).

STRUCTURAL FEATURES AND ORGANOLEPTIC QUALITIES

This study concerns the organoleptic qualities of about 70 aldehydes, i.e. alkanals, alkenals, and alkadienals from C_6 to C_{14}. In these compounds a wide range of organoleptic (olfactive) qualities can be recognised, such as green, herbal, floral, citrusy and fruity. These main characteristics need more specification, such as floral with orris, violet, rose or muguet connotations and citrusy with orange, lemon, mandarin or grapefruit resemblance. Moreover, the concentration at which the compounds are judged is important for the determination of organoleptic quality. On dilution, more characteristics can often be recognised. In concentrated form (over 1%) the aldehydes lower than C_{10} often have penetrating/pungent qualities and the aldehydes over C_{10} possess fatty characteristics. It is preferable to judge the organoleptic qualities in 0·1% concentration (or sometimes even at ppm level) in an inert odourless and tasteless solvent.

With respect to its structural features, every aldehyde can be

TABLE 13
Correlation of Threshold Values of C_9 Aldehydes with Dipole Moments

Aldehyde	TV_{odour} *in water*	TV_{odour} *in air*	*Dipole moment*	TV_{odour} *in min. oil*	$TV_{flavour}$ *in min. oil*
	(*g/10⁹ ml*)		μ	(*ppm*)	
Nonanal	1	30×10^{-3}	2·5	13·5	0·32
trans-2-Nonenal	0·08	$0·5 \times 10^{-3}$	3·5	3·2	0·1
trans-2,*trans*-4-Nonadienal	0·09	$0·25 \times 10^{-3}$	4–4·5	2·5	0·46
trans-2,*cis*-6-Nonadienal	0·01	$0·06 \times 10^{-3}$	3·0	0·01	0·002

divided into the two factors which determine the electronicity and the stereocity of the molecule. The electronicity is mainly determined by the functional group, in relation to the other parts (polar or apolar) of the molecule. The stereocity represents the complete profile of the molecule, e.g. determined by size, shape and chirality [*cis* (*Z*) and *trans* (*E*) geometrical isomerism].

In normal straight chain aldehydes, one can simply recognise the aldehyde function and the alkyl group. The structural features of unsaturated aldehydes are more complicated. For this study we assumed that n-alkanals, *trans*-2-alkenals, and *trans*-2,*trans*-4-alkadienals have, as integral functional groups, —CHO (F_1), —CH=CH—CHO (F_2) and —CH=CH—CH=CH—CHO (F_3), each with their own polarity and stereocity (size and shape). These aldehydes, moreover, have an alkyl group which completes their profiles. In unsaturated aldehydes with isolated double bonds, we regard each double bond as a functional group, at its distinct place. It should be noticed, however, that isolated 'skipped' double bonds (—CH=CH—CH_2—CH=CH—) are more or less functional entities; the same holds true for the double bond similarly separated from the aldehyde function (—CH=CH—CH_2—CHO).

Regarding the organoleptic qualities (odour and flavour) and the structural features of aldehydes in dilution (0·1%), one can subdivide at least six groups of compounds:

(a) green (leaf, violet);
(b) vegetable (cucumber, melon);
(c) fatty (cream, fatty oils, fatty acids);
(d) floral (orris, rose, muguet);
(e) herbal (calamus);
(f) citrusy (orange, lemon, mandarin, grapefruit).

Some indications can be obtained of which structural features characterise a particular organoleptic quality, for instance:

cis-3-hexenal: $C_2H_5CH{=}CHCH_2CHO$: green leaves;
trans-2,*cis*-6-nonadienal: $C_2H_5CH{=}CH(CH_2)_2CH{=}CHCHO$: green-violet leaves;
cis-4,*cis*-7-deca- and -undeca-dienal: $RCH{=}CHCH_2CH{=}CH$-$(CH_2)_2CHO$: herbal (calamus);
nonanal and undecanal: RCHO: floral (orris, rose-fatty);
cis-4-decenal: $RCH{=}CH(CH_2)_2CHO$: floral (muguet).

TABLE 14
Aldehydes with Citrusy/Orange Organoleptic Qualities

Aldehyde	*Citrusy/ orange character/ intensity*	*R*	$F_3^{E/Z}$	F_2	F_1
C_6^{2E}	1	$H(CH_2)_3$		$CH\overset{E}{=}CH$	CHO
$C_6^{2E,4E}$	1	$H(CH_2)_1$	$CH\overset{E}{=}CH$	$CH\overset{E}{=}CH$	CHO
C_8	2	$H(CH_2)_7$			CHO
C_8^{2E}	2	$H(CH_2)_5$		$CH\overset{E}{=}CH$	CHO
$C_8^{2E,4E}$	2	$H(CH_2)_3$	$CH\overset{E}{=}CH$	$CH\overset{E}{=}CH$	CHO
C_{10}	3	$H(CH_2)_9$			CHO
C_{10}^{2E}	3	$H(CH_2)_7$		$CH\overset{E}{=}CH$	CHO
$C_{10}^{2E,4E}$	3	$H(CH_2)_5$	$CH\overset{E}{=}CH$	$CH\overset{E}{=}CH$	CHO
$C_{10}^{2E,4Z}$	4	$H(CH_2)_5$	$CH\overset{Z}{=}CH$	$CH\overset{E}{=}CH$	CHO
C_{11}	1	$H(CH_2)_{10}$			CHO
C_{11}^{2E}	2	$H(CH_2)_8$		$CH\overset{E}{=}CH$	CHO
C_{11}^{5Z}	2	$H(CH_2)_5CH\overset{Z}{=}CH(CH_2)_3$			CHO
C_{11}^{10}	1	$HCH{=}CH(CH_2)_8$			CHO
C_{12}	1	$H(CH_2)_{11}$			CHO
C_{12}^{2E}	2	$H(CH_2)_9$		$CH\overset{E}{=}CH$	CHO
$C_{12}^{2E,4E}$	3	$H(CH_2)_7$	$CH\overset{E}{=}CH$	$CH\overset{E}{=}CH$	CHO
C_{13}	1	$H(CH_2)_{12}$			CHO
C_{13}^{2E}	2	$H(CH_2)_{10}$		$CH\overset{E}{=}CH$	CHO
Average	1·95	$(CH_2)n = 6{\cdot}62$	$F = 0{\cdot}38$	$F_2 = 0{\cdot}32$	$F_1 = 0{\cdot}30$

The question arises, are there particular features for a particular organoleptic quality? Some simple indications are given below:

(a) alkyl chain over C_{10}	fatty
(b) even and odd C number	citrusy and floral
(c) 3-*cis* and 6-*cis* double bonds	green
(d) 4-*cis* double bonds	floral (muguet)
(e) 4-*cis*,7-*cis* 'skipped' double bonds	herbal (calamus)

In the following scheme the organoleptic qualities of C_{10} aldehydes are demonstrated:

n-decanal (concentrated): fatty, citrusy;
n-decanal, *trans*-2-decenal, *trans*-2,4-*cis*/*trans*-decadienal (diluted): citrusy (strong orange);
cis-3-decenal: green, cucumber, melon;
cis-4-decenal: floral (muguet);
trans-2,*cis*-5-decadienal: green (violet leaf);
cis-4,*cis*-7-decadienal: herbal (calamus);
9-decenal: floral (rose-fatty);
cis-4,*cis*-7,9-decatrienal: floral, citrusy, herbal.

We evaluated the aldehydes C_6 to C_{14}, which are depicted in Table 14, for their citrusy/orange character. The length of the alkyl chain and the importance of the functionalities F_1, F_2 and F_3 are demonstrated. From this table it may probably be concluded that the *cis*,*trans* isomers of 2,4-alkadienals with 8, 10 and 12 carbon atoms in dilution may have the most pronounced orange quality.

FINAL REMARKS AND CONCLUSIONS

A possibility of verification of threshold values of homologous series in different media by calculating the partition coefficient (K) and $\log K$, which shows a linear relation with the number of carbon atoms, has been demonstrated. Some indication of correlation of threshold values of aldehydes with physicochemical parameters has been derived. The most pronounced correlation was found with molecular weights [interrelated to molar volume, boiling point, molar refractivity and partition coefficient (P)] and melting points.

The influence of the position and geometry of double bonds, in aldehydes, on the organoleptic quality has been demonstrated. The structural features of aldehydes with citrusy (orange) character were studied in more detail.

REFERENCES

1. Straten, S. van, Maarse, H., de Beauverser, J. C. and Visscher, C. A. (1983). *Volatile Compounds in Food: Qualitative Data*, 5th edition, Division for Nutrition and Food Research TNO, Zeist (and supplements).
2. Arctander, S. (1969). *Perfume and Flavor Chemicals,* published by the author, Montclair, N.J. (USA).
3. Bedoukian, P. Z. (1967). *Perfumery and Flavoring Synthetics,* **19**, Elsevier, Amsterdam; (1985) *Perf. and Flav.*, **10**, 75.
4. Guenther, E. (1949). *The Essential Oils*, Vol. 2, Van Nostrand, New York, 305–26.
5. Gildemeister, E. and Hoffmann, F. (1963). *Die Aetherische Oele*, Band IIIc, Akademie-Verlag, Berlin, 20–50.
6. Boix Champs, A. (1985). *Perf. and Flav.*, **10**, 6–10.
7. Lier, F. P. van, Linde, L. M. van der and Weerdt, A. J. A. van der (1986). In: *The Essential Oils,* Ed. Bruncke, E. J., in press.
8. Meijboom, J. W. and Jongenotter, G. A. (1981). *J. Am. Oil Chem. Soc.*, **58**, 680–2.
9. Meilgaard, M. C. (1975). In: *Geruch- und Geschmackstoffe,* Ed. Drawert, F., Verlag Hans Carl, Nürnberg, 233–8.
10. Lea, C. H. and Swoboda, P. A. T. (1958). *Chem. and Ind.*, **46**, 1289–90.
11. Day, E. A., Lillard, D. A. and Montgomery, M. W. (1963). *J. Dairy Sci.*, **46**, 291.
12. Siek, T. J., Albin, I. A., Sather, L. A. and Lindsay, R. C. (1969). *J. Food Sci.*, **34**, 265.
13. Meijboom, P. W. (1964). *J. Am. Oil Chem. Soc.*, **41**, 326–8.
14. Meijboom, P. W. and Jongenotter, G. A. (1971). *J. Am. Oil Chem. Soc.*, **48**, 143.
15. Meijboom, P. W. and Jongenotter, G. A. (1975). *Fette Seifen Anstrichm.*, **77**, 135.
16. Dixon, M. D. and Hammond, E. G. (1984). *J. Am. Oil Chem. Soc.*, **61**(9), 1452–6.
17. Guadagni, D. G., Buttery, R. G. and Orano, S. (1963). *J. Sci. Food Agric.*, **14**, 761–5.
18. Guadagni, D. G., Buttery, R. G. and Turnbaugh, Jean G. (1972). *J. Sci. Food Agric.*, **23**, 1435–44.
19. Buttery, R. G., Ling, L. C. and Guadagni, D. G. (1969). *J. Agric. Food Chem.*, **17**(2), 385–9.
20. Buttery, R. G., Guadagni, D. G. and Ling, L. C. (1971). *J. Agric. Food Chem.*, **21**(2), 198–201.

21. TERANISHI, R., BUTTERY, R. G. and GUADAGNI, D. G. (1974). *Ann. NY Acad. Sci.*, **237**, 209–16.
22. AMOORE, J. E. and BUTTERY, R. G. (1978). *Chem. Sens. Flavor,* **3**, 57–71.
23. HALL, G. and ANDERSSON, J. (1983). *Lebensm. Wiss. Technol.*, **16**, 354–61, 362–6.
24. CHRISTOPH, N. and DRAWERT, F. (1985). In: *Topics in Flavour Research*, Ed. Eichhorn, H., Marzburg-Hangenborn, 59–77.
25. ROTHE, M., WOLM, G., TUNGER, L. and SIEBERT, H.-J. (1972). *Nährung,* **16**, 483.
26. SCHNABEL, K.-O. (1982). Dissertation, TU München.
27. BACKMANN, E. L. (1917). *J. Physiol. Path. Gen.*, **17**, 1–4.
28. GRAMSHAW, J. W. and WILLIAMS, D. R. (1981). In: *Flavour '81*, Ed. Schreier, P., de Gruyter, Berlin/New York, 165–78.
29. GEMERT, L. J. VAN and NETTENBREIJER, A. H. (1977). *Compilation of Odour Threshold Values in Air and Water,* Central Institute for Nutrition and Food Research TNO, Zeist, The Netherlands, and Nat. Inst. for Water Supply, Voorburg, The Netherlands.
30. GEMERT, L. J. VAN (1984). *Compilation of Odour Threshold Values in Air,* Suppl. V, Report no. A84.220, CIVO-Institutes TNO, Zeist, The Netherlands.
31. KUMLER, W. D. and FOHLEN, G. M. (1945). *J. Am. Chem. Soc.*, **67**, 437–41.
32. MUELLER, A. (1957). *Int. Riechstoff Kodex,* Erg. Bd. 1, Huethig Verlag, Heidelberg, 241.

3

Modern Sensory Analysis and the Flavour Industry

ANTHONY A. WILLIAMS

AFRC Institute of Food Research, Shinfield, Reading, UK

ABSTRACT

Although current descriptive procedures and the production of perceptual spaces showing the interrelationships between samples to some extent mimic what flavourists have been doing for years, they do provide a means of rationalising and clarifying their thought processes, thus enabling them to get a better insight into the way they can modify and improve products. This paper discusses some recent developments in sensory analysis, in particular those associated with conventional and free choice profiling and similarity scaling, and shows how these, coupled with modern statistical techniques, can be used to help the flavourist. Inspection of the sample spaces which can be derived from any of the above procedures shows how samples interrelate in sensory terms. By the use of regression models, dimensions or vectors can be placed through this space to indicate how the sensory characteristics of a finished product may be altered by changing the composition of a particular flavouring ingredient blending or, in the case of a single component, by structural variations. Linking this information with target ideals, one can also determine how such variation affects the suitability of a flavour for a particular application. Experience with foods and beverages in general indicates that individuals perceive foods and their weight characteristics differently when evaluating their relative acceptance. When flavouring foods, too, the tolerance limits of flavour characteristics which consumers are prepared to accept are equally as important as defining the characteristics of a target product. The implications of these factors when flavouring foods, and how the introduction of the microcomputer into the sensory laboratory is opening up new ways of evaluating foods (particularly when trying to determine optimum blends of ingredients and tolerance limits), are discussed.

INTRODUCTION

Twenty years ago scientific sensory evaluation was largely difference testing and ranking; profiling was in its infancy.[1] Since then there have been many advances both in respect to methodology and how information is interpreted,[2,3] much of this hinging around the concept of a multidimensional perceptual space which reflects in terms of distances the way people perceive samples and their interrelationships.

The derivation of sample spaces and the way the sensory analyst examines and interprets them to some extent mimics the thought processes of the flavourist, the perfumer and commodity experts, such as master wine or coffee blenders. All these people aim to describe products and determine how they differ from some preconceived objective. Their experience tells them what attributes are missing or need modifying and as a consequence what notes need to be added and what to add to give these notes. Integration of this expert knowledge with the concepts of modern sensory analysis, either by the flavourist on his own or by integrating his or her expertise with that of a trained panel, should enable such thought processes to be rationalised and recorded. The use of such procedures would enable the flavourists not only to get a better and more logical insight into the way they can modify and improve products but also provide them with a useful vehicle for communicating with other flavourists or experts and of course customers in the food industry, technical people and administrators.

THE PRODUCTION OF A SAMPLE OR PERCEPTUAL SPACE

As far as the flavour characteristics of a product are concerned, a sample or perceptual space is very conveniently produced by some form of descriptive assessment such as conventional[4] or free choice profiling.[5] Simple difference scaling procedures[6] can produce similar

TABLE 1
Aroma Attributes Used to Assess Bordeaux Wines

Berry (blackberry/raspberry)	Raisin
Blackcurrant (canned/Ribena)	Soy/Marmite
Synthetic fruit	Vanilla
Green bean/green olive	Phenolic/spicy
Black pepper	Ethanol

information in respect to the interrelationships of samples but, as no descriptive terms are used in deriving the space, its interpretation is difficult and inevitably relies on external information.

In profiling, a list of descriptors describing the various characteristics of the products in question is developed. These may be complete foods, plant extracts, flavours or simple flavouring compounds. The derived list may be a set of standard terms with meaning agreed between the assessors as in conventional profiling[4] or lists of terms which are peculiar to each individual assessor as in free choice profiling.[5] In this latter case each assessor may use as many or as few terms as he wishes and there is no need for any individual or panel leader to know what any other person means by his or her descriptors.

Typical terms developed for a conventional profile evaluation of wines are given in Table 1,[7] and a selection of terms which would be of potential value in the free choice profiling of flavouring compounds such as thiazines is given in Table 2.

TABLE 2
Examples of Terms Potentially Usable in the Free Choice Profiling of Thiazines

Green	Nutty
Peppery	Roasted peanut
	Hazel nut
Vegative	
Cabbage like	Meaty
Sulphury	Thiamine like
Oniony	Yeasty
Burnt	
Phenolic	Rancid
Spicy	Fatty
Stale	
Metallic	

When operating, these terms are incorporated into a common score sheet in the case of conventional profiling or individual score sheets in the case of free choice profiling. Each product is evaluated by each assessor scoring each of the attributes on the amount present in a particular sample, e.g. 0 if absent and 5 if present to a high level. In this way each individual builds up a picture of each of the products being examined in terms of words and numbers, information which in itself provides a valuable record.

If a range of samples is being evaluated, such information can also be used to produce a sample or perceptual space, either in respect to individuals or for the panel as a whole. If each of the adjectives being scored is represented by a dimension in a multidimensional space, each sample can be given a unique position in that space depending on the way it is being scored by an individual or panel.[3,8,9] The distance between samples in such a space clearly shows their similarities and differences.

If one uses individual data, these spaces reflect how the individuals perceive the interrelationships between samples and are independent of how a person has used terms (whether free choice profiling or otherwise) or scales. If one is to compare such information between two flavourists or rationalise it over a set of panellists, it is necessary

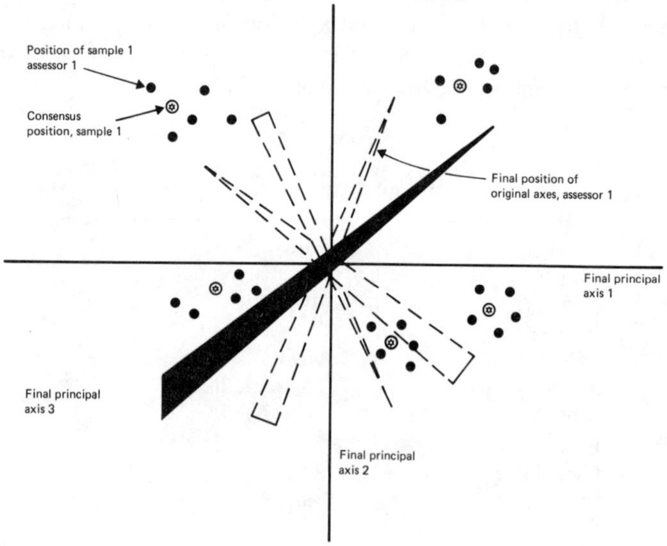

FIG. 1. Matching of individual sample spaces to produce a consensus configuration.

to determine just how similar these spaces are. By centring individual spaces, placing one on top of the other and applying various rotational stretching and shrinking techniques,[10] it is possible to attempt to match each of these individual spaces so that all similar samples are as close together as possible (Fig. 1). The important fact to note is that during these operations individual intersample relationships are unaffected: they might be rotated and stretched or shrunk as a whole, but the actual relationship of one sample to another remains unchanged. The two operations of centring the sample spaces and stretching and shrinking permit one to allow for variation in the way people score attributes, whereas that of rotation allows one to take into account different uses of terms in the case of conventional profiling and to match completely different terms in the case of free choice profiling. By taking the centroid of each of the sample clusters it is possible to produce a consensus sample configuration (Fig. 1).

When describing foods or flavours, however, people inevitably use more than three descriptors, so the problem of visualising the multidimensional space arises. Using the appropriate statistical technique such as principal component, factor or canonical variate

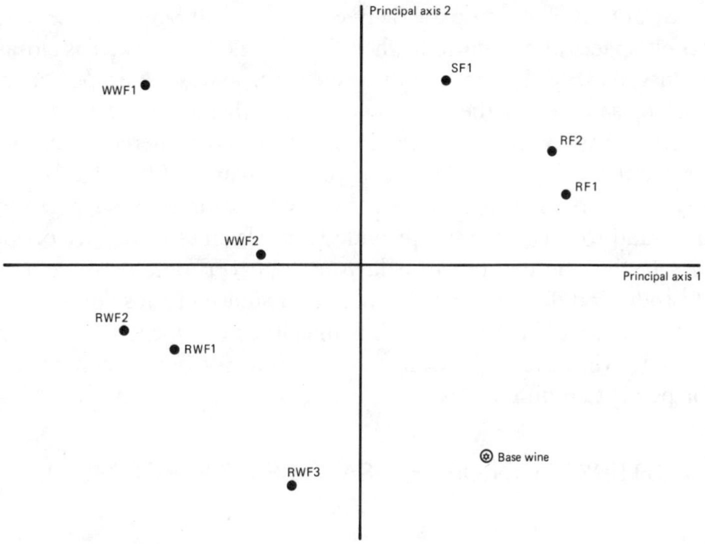

FIG. 2. Simulated principal component analysis of flavoured wines. WWF = white wine flavour; RWF = red wine flavour; SF = sherry flavour; RF = raisin flavour.

analysis,[11] a line or plane may be taken through such a space and, by projecting all the samples on to it, the original multidimensional space can be reduced to one of one or two dimensions which can be visualised. The most common and useful of these approaches is principal component analysis in which lines and planes are chosen to be orthogonal to one another and account for decreasing amounts of variation. It therefore produces a low dimensional space accounting for the maximum variation between the samples being examined. By projecting the original axes on which the sample space was based onto these new dimensions, the latter can be interpreted in terms of the original vocabularies used to assess the samples. Figure 2 provides an example of how such procedures might be used to determine and illustrate the relative effects of adding different flavourings to a base wine.

SIGNIFICANCE OF VARIATION IN INDIVIDUAL PERCEPTUAL SPACES

By examining the nature of individual assessors' sample spaces after rotation, the sorts of terms people are using to describe similar underlying dimensions in a set of products can be determined. More important, by examining the degree of agreement between assessors' sample spaces after rotation, when they should be matched as closely as possible (Fig. 1), one can discover if the assessors are perceiving a set of samples in the same way or whether they are perceiving different sensory characteristics in the product (not merely describing them differently). Apart from helping to clarify differences in the way assessors or flavourists are perceiving samples, if such groups are found to exist, it could provide valuable clues as to why people vary in their acceptance of a flavour. If two people perceive a set of products differently their relative acceptance of those products is also likely to differ. Even if such people like the same thing, different flavours will have to be designed for each assessor to compensate for perceptual differences.

UTILISATION OF THE SAMPLE OF PERCEPTUAL SPACE

Apart from identifying groups of people who perceive flavour differently, the perceptual space model is invaluable for relating to external information, e.g. the acceptance of final flavoured products, change in the length of a side chain if one is looking at the characteristics of a homologous series of compounds or processing variable if one is interested in the flavour character of a processed product. One can collect hedonic information from consumers, experts or customers and by using various regression models place dimensions through the sample space which relate to acceptance.[3,8,9,12,13] If the customers' or consumers' ideal lies within the space rather than on its extremes, surface response methodology can also be used to find out its characteristics. Alternatively, assessors can be asked to imagine ideal products or to score attributes in respect to ideal values. As a consequence, ideal samples can be characterised in relation to the products being examined and points placed in the perceptual space relating to them.

All these approaches enable the flavourist to see just what needs to be done to any particular sample to improve it, as well as discover what sensory characteristics are most important to a particular assessor when deciding preferences. When dealing with a processed product, the dimensions can also be placed through the space which relates to production variables. By comparing such dimensions with the acceptance dimensions it is possible to discover whether altering a particular part of a process will have any effect on the acceptability of the sensory properties of the product.

In the rather simplified case discussed above, one dimension is shown relating to acceptability. Just as people are different and perceive different things in a product, they may also like different things and put different emphasis on different aspects of the product when deciding on its acceptability. This may be the case even when they perceive it in the same way. If, within a sample space, individual data relating to acceptance are explored, it becomes apparent that each individual has his or her own ideas about what he or she should be eating or drinking. Vectors relating to acceptability for one individual may therefore be different from those of another. It might be expected, however, that such vectors will cluster so that eventually it becomes possible to identify further sub-populations who not only perceive products in the same way but also rate them similarly and have similar concepts of what they should be consuming.

Taking the example of the flavoured wines (Fig. 3), different vectors may well have been encountered (a) depending on whether

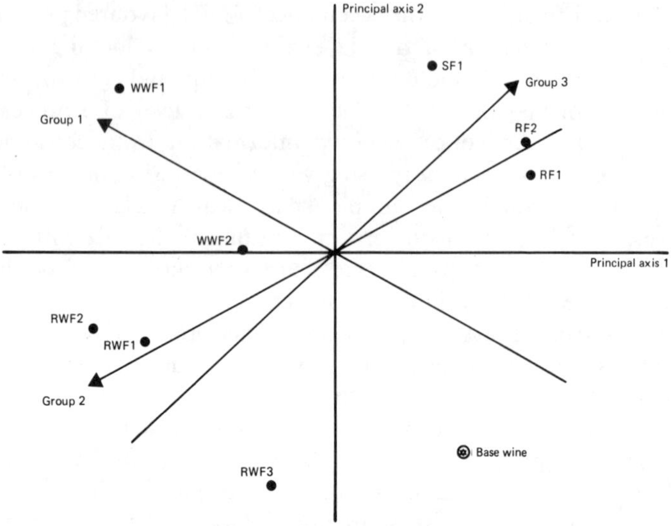

FIG. 3. Vectors relating to acceptance of flavoured wines. For identities of samples see Fig. 2. Group 1 = assessors who like white wine flavour; Group 2 = assessors who like red wine flavour; Group 3 = assessors who like sherry/raisin flavour.

assessors were primarily white or red wine drinkers or, in the case of this example, if they were red wine drinkers depending on whether they preferred (b) raisiny or (c) more grapey type wines. As well as producing vectors, surface response approaches could also enable individual ideals to be pinpointed (Fig. 4), particularly if they fall within the derived space.

Having identified significant vectors or characterised ideal products, this information can then be related to the composition of the flavour in chemical terms,[14] information such as that obtained by gas chromatography or gas chromatography/mass spectrometry. Whereas maximum likelihood modelling provides useful pictures when trying to unravel hedonic information, principal component regression and partial least squares regression[15] seem to offer more potential when relating sensory and chemical data. This latter procedure not only finds the best predictive relationship but also takes into account the degree of variation in the parameter being measured.

When considering ideal products and trying to characterise the flavour which best satisfies it, the tolerance of individuals to change

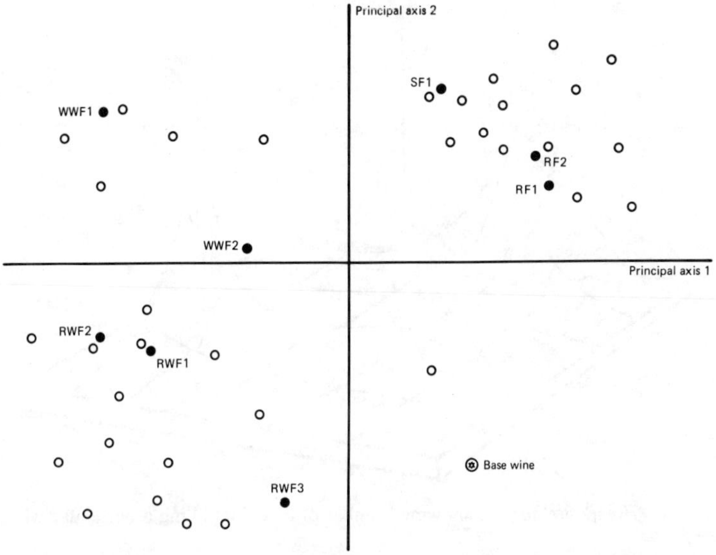

FIG. 4. Individual assessors' ideal wines superimposed on flavoured wine space. For identities of samples see Fig. 2.

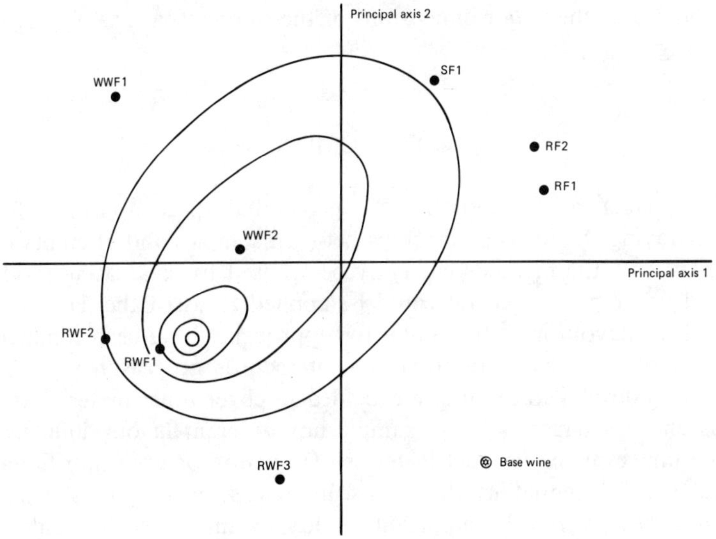

FIG. 5. Contour plot showing fall-off in acceptance for one assessor. For identities of samples see Fig. 2.

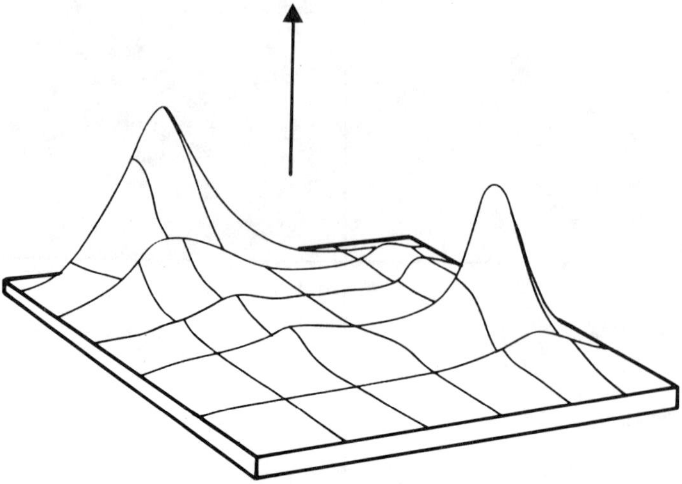

FIG. 6. Response surface showing number of assessors liking a particular wine.

in composition is equally as important as the characteristics of the ideal itself. Such information may be presented in the form of a contour plot as shown in Fig. 5 or in the form of a surface response model over the major dimensions of the sample space as indicated in Fig. 6.

CONCLUSIONS

This paper briefly describes the perceptual space approach for portraying the interrelationships between samples and attempts to show how the methods currently being used for evaluating foods and beverages in general could be applied to assist the flavourist and the flavour industry. So far the approaches have been confined primarily to sensory information but there is no reason why the models developed cannot be extended to cover other material and psychological factors,[13] for example, how different flavours influence the image of the product in terms of function or even nutritional value and potential health benefits. In the case of fragrances, where imagery is extremely important, ability to analyse such nebulous factors as how people perceive the product and what they believe it is doing for them is probably far more important than the sensory characteristics of the fragrance itself. Irrespective of the commodity, however, understanding why customers purchase what they do is important to all people in the flavour industry. Although there are still many ways in which sensory models can be improved, some powerful techniques already exist giving the scientist insight into the factors affecting food acceptance. The flavour industry could benefit from making use of them and developing them further for their own use.

REFERENCES

1. AMERINE, M. A., PANGBORN, R. M. and ROESSLER, E. B. (1965). *Principles of the Sensory Evaluation of Foods*, Academic Press, London.
2. WILLIAMS, A. A. (1983). *Chem. Ind.*, 740–5.
3. WILLIAMS, A. A. (1985). *Food Flavourings, Ingredients, Processing, Packaging*, **59**, 28–31.
4. WILLIAMS, A. A. (1975). *J. Sci. Food Agric.*, **26**, 567–82.
5. WILLIAMS, A. A. and LANGRON, S. P. (1984). *J. Sci. Food Agric.*, **35**, 558–68.
6. MCFIE, H. J. H. and THOMSON, D. M. H. (1984). In: *The Sensory Analysis of Foods*, Ed. Piggott, J. R., Applied Science, London, 351–75.
7. NOBLE, A. C., WILLIAMS, A. A. and LANGRON, S. P. (1984). *J. Sci. Food Agric.*, **35**, 88–98.
8. WILLIAMS, A. A. (1984). In: *Tartrates and Concentrates, 8th Wine Subject Day*, Ed. Beech, F. W., Long Ashton Research Station, 3–12.
9. WILLIAMS, A. A. and ARNOLD, G. M. (1986). In: *Proceedings of the First International Symposium on Food and Health*, Bioscience Endprint, Geneva (in press).
10. LANGRON, S. P. (1983). In: *Sensory Quality in Foods and Beverages: Definition, Measurement and Control*, Eds Williams, A. A. and Atkinson, R. K., Ellis Horwood, Chichester, 89–95.
11. CHATFIELD, C. and COLLINS, A. J. (1980). *Introduction to Multivariate Analysis*, Chapman and Hall, London, New York, 246.
12. SHIFFMAN, S. S., REYNOLDS, M. L. and YOUNG, F. W. (1981). *An Introduction to Multidimensional Scaling Theory Methods and Application*, Academic Press, London, 413.
13. WILLIAMS, A. A. (1986). In: *EEC Workshop on Evaluating the Acceptability of Agricultural Commodities*, in press.
14. WILLIAMS, A. A., ROGERS, C. and NOBLE, A. C. (1984). In: *Flavour Research of Alcoholic Beverages*, Eds Nykanen, L. and Lehtonen, P., Foundation for Biotechnical and Industrial Fermentation Research, **3**, 235–53.
15. WOLD, S., ALBANO, C., DUNN III, W. J., ESBENSEN, K., HELLBERG, S., JOHANSSON, E. and SJÖSTROM, M. (1983). In: *Food Research and Data Analysis*, Eds Martens, H. and Russwurm, H., Applied Science, London, 147–88.

4

Recent Developments in Umami Research

YOSHI-HISA SUGITA
Society for Research on Umami Taste, Tokyo, Japan

ABSTRACT

Umami is defined as the taste of a group of amino acids and nucleotides such as glutamates, inosinates and guanylates. The Society for Research on Umami Taste (SRUT) was started in 1982 in Japan as an interdisciplinary study group consisting of professors in neuroscience, physiology, nutrition, food science, psychology and chemistry. After having had a successful symposium in Tokyo in 1984, SRUT held an international symposium in Hawaii in 1985. Lots of novel data accumulated on this taste may be sufficient to characterise umami as an independent fundamental taste along with the four basic tastes of Henning. An update summary of the development of research in this field will include the following: (1) occurrence in nature: umami and food acceptability; (2) psychometric study: umami and the other basic tastes, synergism in glutamate and nucleotide, omission test, flavour enhancer; (3) neurophysiology: receptor; (4) nutrition/behaviour: cafeteria study, sodium chloride/protein/glutamate, taste aversion test, infant perception.

INTRODUCTION

Umami is defined as the taste of monosodium glutamate (MSG) and 5′-nucleotides such as 5′-inosinate (IMP) and 5′-guanylate (GMP). The chemical properties of umami substances have been extensively studied by Japanese scientists and have been well elaborated. In contrast, the physiology of umami—its sensory evaluation, receptor mechanisms, and its role in food tastes—has not been thoroughly investigated, and many questions remain unresolved. During the past several decades, the number of researchers studying umami has greatly increased, and important experimental data have accumulated. Most of these results were presented at the International Symposium on Umami which was held in Kauai, Hawaii, from 8th to 10th October, 1985. Physiological studies on the taste of umami have been carried out from various approaches, i.e. neurophysiology, biochemistry, psychophysics, food chemistry, and nutrition. It is not possible to review all topics, but the major findings that were presented that relate directly to umami are introduced here.

TASTE RECEPTOR MECHANISM OF UMAMI

The biological significance of the chemoreception of amino acids has been reviewed. Amino acids are one of the most important nutrients and are potent attractants for many living organisms; thus, organisms as diverse as bacteria, protozoa, annelids, gastropods, crustaceans, insects, fish, amphibians and mammals all have, in common, chemoreceptors sensitive to amino acids. Treatment of carp olfactory epithelium, bullfrog tongue, and rat tongue with pronase E obliterates responses to amino acids in these animals. Therefore, receptor molecules for amino acids, including glutamate (Glu) have been definitively shown to be proteins. Cross-adaptation experiments, utilising various combinations of amino acids in olfactory and taste systems of the carp, explored the receptor sites for various amino acids. Results showed that there are multiple types of receptors corresponding to different amino acids and that the receptor for Glu is different from those for other amino acids. Evidence suggesting that there is a specific receptor protein for Glu in chemoreceptor membranes is consistent with reports on receptor potentials in bacteria.

Single-unit nerve responses to Glu and 5′-nucleotides have been recorded in the taste pore, the chorda tympani nerve, the glossopharyngeal nerve, the geniculate ganglion, the petrosal ganglion and cortical neurons in the rat, hamster, mouse, cat, goat, and dog. Results (Fig. 1) varied somewhat, depending on the nerve fibre recorded and the species of experimental animal studied. For

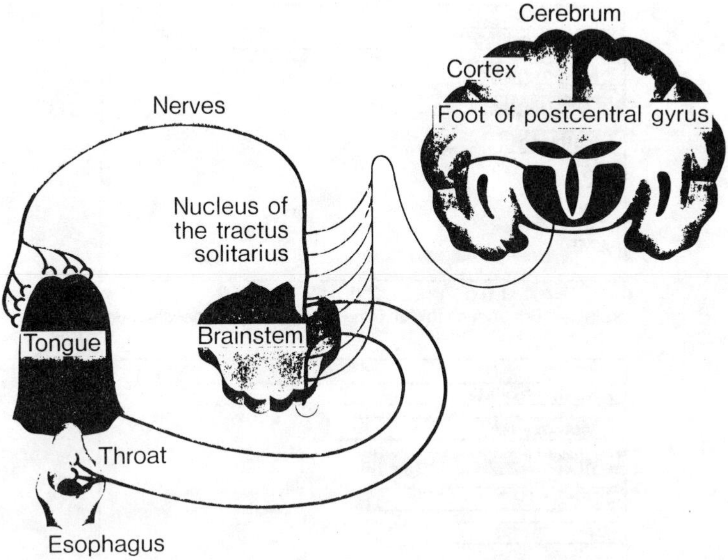

FIG. 1. Signal transmission of taste to the brain.

example, although some units of the cat geniculate ganglion responded strongly to nucleotides, these same units failed to respond to MSG. In hamster, MSG-sensitive units of cortical neurons also responded to NaCl; however, the responses of MSG-sensitive units recorded within the hamster taste pore were independent of the responses to any of the four basic taste stimuli. Some of the MSG-sensitive units of the glossopharyngeal nerve in the mouse responded only slightly to NaCl or to sucrose. Findings from behavioural studies, on responses to various taste stimuli in mice, are consistent with the electrophysiological data.

The mechanism of the taste synergism between Glu and nucleotides has been examined both electrophysiologically and biochemically. Recordings from the rat chorda tympani nerve demonstrate the remarkable synergism (Fig. 2) between MSG and purine-based 5′-nucleotides, which is consistent with and corroborates the psychophysical data. Analysis of the data from the rat has suggested that the presence of 5′-nucleotides induces an increase in the affinity (i.e. strength of binding) of Glu for the receptor sites while not increasing the amount of Glu bound. However, measurements of binding of L-[^{3}H]glutamate to bovine circumvallate papillae showed

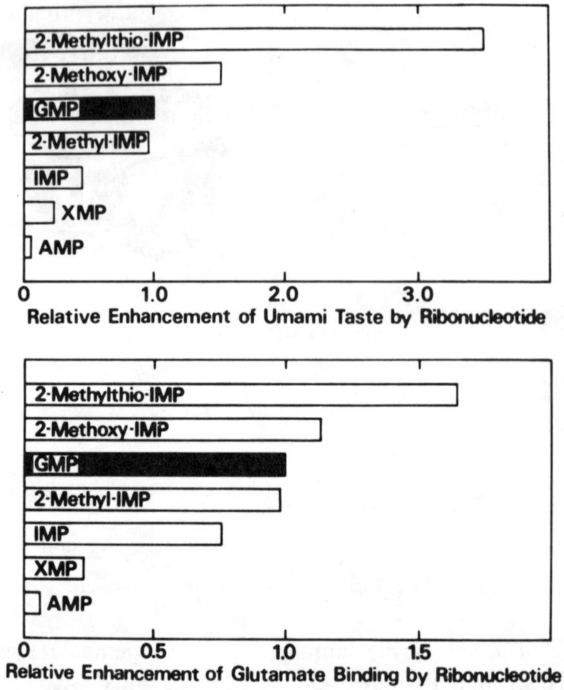

FIG. 2. Ratio of 'umami' taste enhanced by ribonucleotides.

that, in this system, the presence of nucleotides caused a remarkable increase in the maximal capacity of Glu binding to the papillae, without affecting binding affinity. The results were explained in terms of allosteric regulation.

Despite the remarkable synergism between MSG and nucleotides with regard to umami, MSG has no enhancing effect on the responses of the rat chorda tympani to the four basic taste stimuli represented by NaCl, HCl, sucrose, and quinine. This finding, which is consistent with the psychophysical data, suggests that the flavour potentiation resulting from the addition of MSG to certain foods is not brought about by an effect on the intensity of the four basic tastes.

FOOD AND TASTE

Researchers have investigated American and Japanese subjects' conceptualisations and descriptions of umami taste. A taste description test demonstrated differences in responses to MSG taste between monolingual American and Japanese subjects. The most common description given by Americans was 'salty', but this was not a majority opinion, and many other descriptions were given. Japanese subjects, especially trained taste panellists and employees of glutamate producers, tended to use the term 'umami' consistently. (These results remind us of the difficulty in describing the colour of an apple or a tomato without using the term 'red'.) The importance of establishing a universal concept of umami has been emphasised (Fig. 3).

Several multidimensional scaling analyses conducted separately in the United States and in Japan indicated that the taste of MSG is definitely located outside of and apart from the tetrahedron of the traditional four basic tastes, and that umami necessitates construction of another dimension independent of the four 'basic' tastes (Fig. 4).

Another multidimensional analysis has shown umami to be present in the tastes of natural foods. Broths made from animal and fish stock (beef, pork, etc.) fell outside the area spanned by the four basic tastes, but lay close to umami. In contrast, broths made from vegetables were scattered about the five taste areas; however,

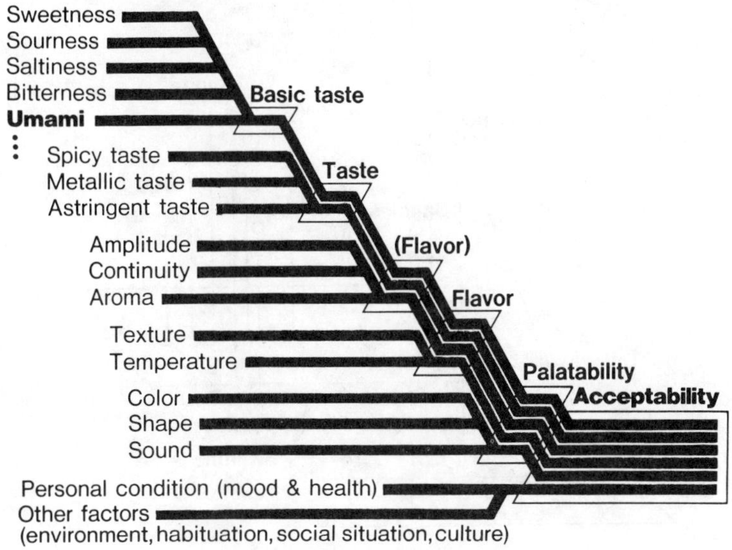

FIG. 3. Food acceptability and umami.

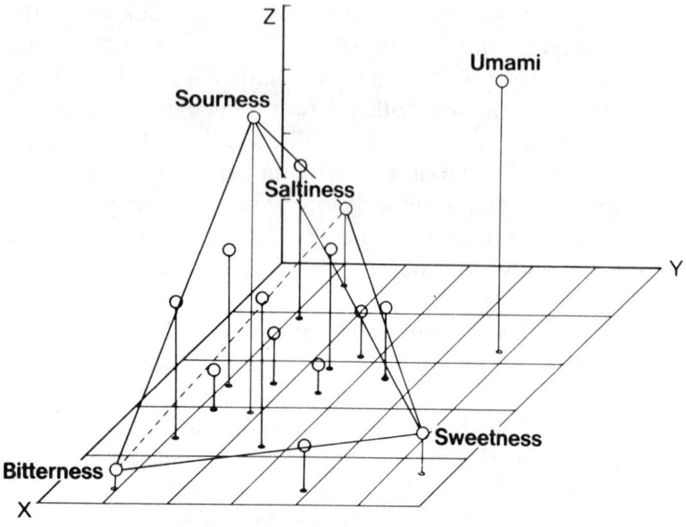

FIG. 4. Uniqueness of umami as one of the basic tastes.

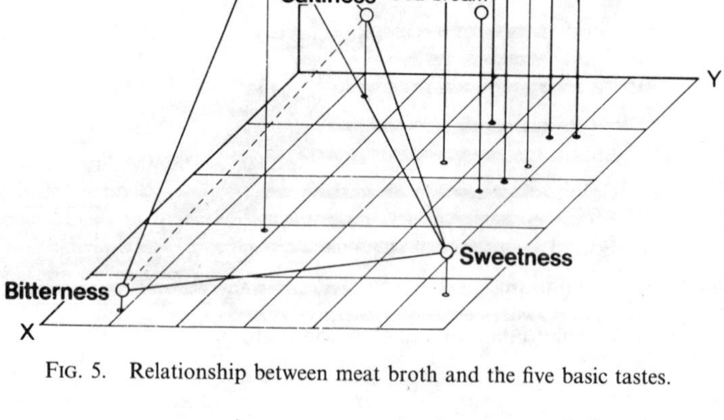

FIG. 5. Relationship between meat broth and the five basic tastes.

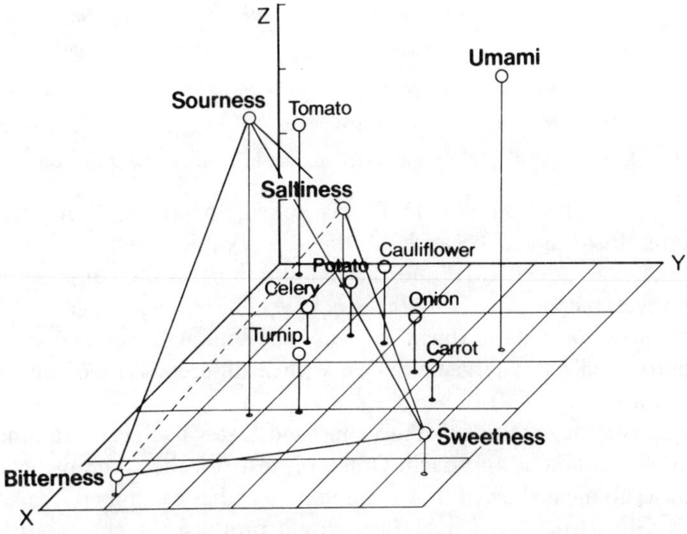

FIG. 6. Relationship between vegetable broth and the five basic tastes.

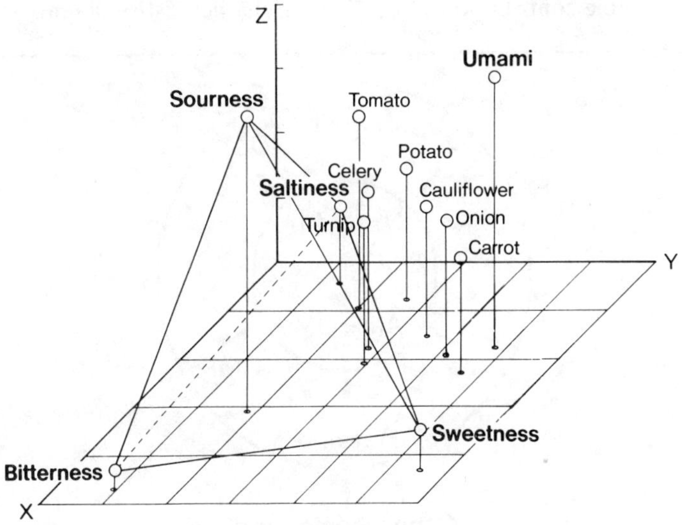

FIG. 7. Relationship between vegetable broth and the five basic tastes (+IMP 0·005%).

Tau Asp Thr Ser Sar Pro Glu Gly Ala AABA Succinic acid NaCl IMP
Val Na_2HPO_4 Ile Leu Tyr Phe Orn Met Inosine Guanine Cytosine KCl
Lys His 3-MeHis Trp Arg CMP AMP GMP Lactic acid NaH_2PO_4 Glucose
Ribose ADP Adenine Adenosine Hypoxanthine Betaine TMAO Homarine · HCl

FIG. 8. Composition of reconstructed extract of boiled snow crab meat.

when a small amount of IMP was added, the tastes approached umami taste space because of the remarkable synergistic effect between the added IMP and the glutamate naturally contained in the vegetable stocks. Thus the stocks examined were regarded to have *dominant* umami (meat) or *potential* umami (vegetable) which is actualised by synergistic action with a different type of umami substance (Figs 5–7).

The role of umami substances in food tastes has been examined also by a different approach. Omission tests on the components of snow crab meat showed that a simplified synthetic system (Gly, Ala, Arg, Glu, IMP, NaCl, KH_2PO_4) could produce the characteristic crab flavour (Figs 8 and 9). When umami constituents contained in the simplified system were eliminated, not only the continuity, complexity, fullness, and mildness disappeared, but even the characteristic flavour of crab decreased. Thus umami substances made a remarkable contribution to the flavour characteristics of crab.

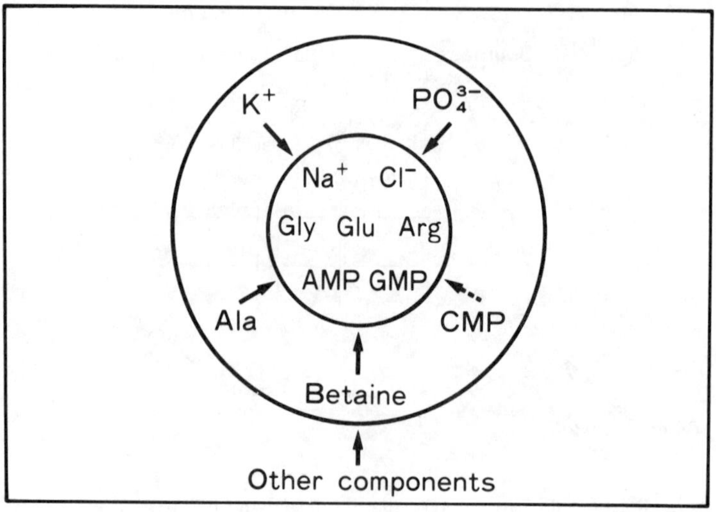

FIG. 9. Composition of simplified model of extract of boiled snow crab meat.

The role of umami taste in meat flavour has also been studied. The strength of umami and brothy tastes is higher in pork and in chicken than in beef, which contains less MSG. The addition of MSG or MSG plus IMP to broths enhanced the umami and brothy tastes, and showed that umami plays an important role in meat taste. Moreover, the organoleptic properties of umami constituents in various food systems have been shown in analyses of many proteins from different sources.

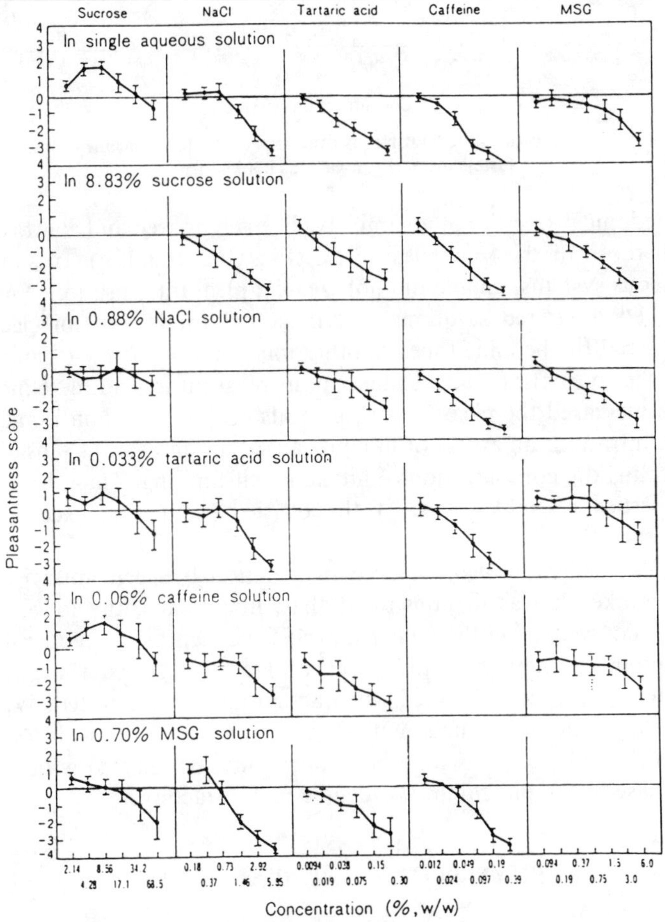

FIG. 10a. Mean pleasantness ratings as functions of the log concentrations of five taste substances in distilled water and in the respective taste solutions.

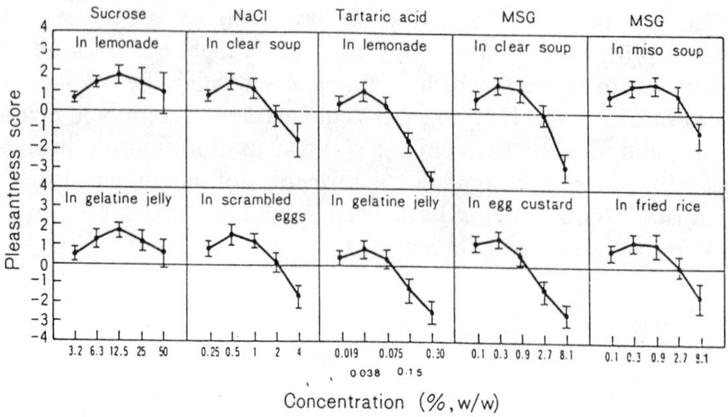

FIG. 10b. Mean pleasantness ratings as functions of the log concentrations of four taste substances in various cooked foods.

Hedonic properties of umami have been compared and contrasted with those of the four basic tastes (Figs 10a and 10b). In simple aqueous systems, umami did not create a pleasant sensation. Yet in selected flavoured solutions and in actual foods, umami clearly enhanced the hedonic tone. Another important property of umami was the concentration dependence of its pleasantness rating. Umami thus increased the pleasantness of foods only in a certain range of concentration; an excess of umami caused an unpleasant sensation and thus the concentration of intake is self-limiting. These hedonic properties resemble those of the other basic tastes, except for sweetness.

This discussion also explored the relation between umami and Na intake. It was demonstrated that, in humans, the lower the NaCl concentration the higher the MSG concentration required to maintain the high palatability score of a clear soup. These phenomena were also observed when actual meals were tested. When a certain amount of umami substances was added to meals, NaCl could be reduced; thus total Na intake could be reduced without a decrease in the palatability or degree of satisfaction.

PHYSIOLOGY AND BEHAVIOUR

The perinatal human infant can communicate feelings and emotions by a repertoire of non-verbal communications consisting of facial

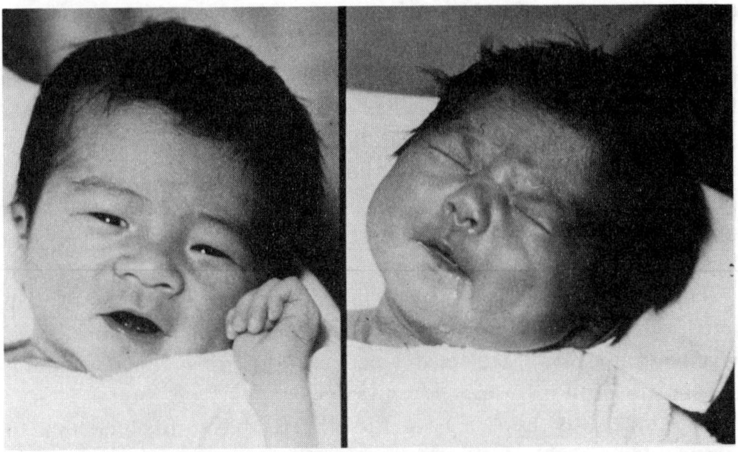

FIG. 11. Facial expressions of new-born babies on being given taste substances.

expressions and features. When a sucrose solution is delivered into its mouth, an infant responds with characteristic pleased and satisfied facial and oral expressions (Figs 11 and 12). When sour or bitter tastes or unflavoured vegetable soup are given, infants respond with features indicating displeasure and dissatisfaction. If glutamate is

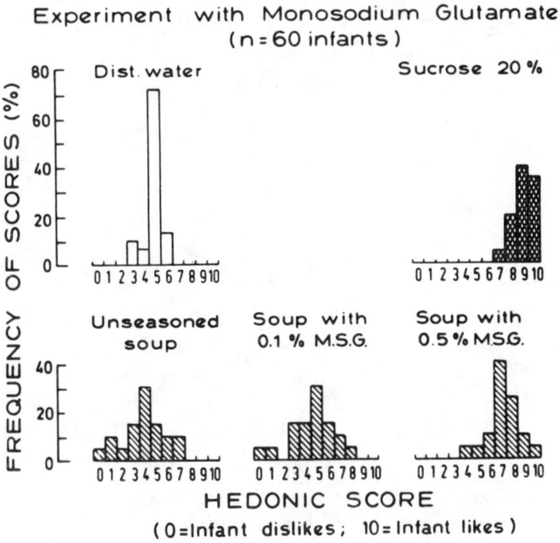

FIG. 12. Facial motion components activated by tastants.

added to the unflavoured vegetable soup, however, infants not only detect its presence but also indicate, with their facial expression, enjoyment of the taste in a manner distinct from their responses to sweet solutions. Similar behavioural manifestations have been observed in rats given dietary glutamate.

Both protein-malnourished infants and well-nourished infants preferred glutamate-seasoned soup to unseasoned soup; however, only malnourished infants preferred soup prepared with casein hydrolysate, with its bitter notes.

Taste preferences in rats were also altered by changes in protein nutritional status. Thus in protein deficiency, sodium chloride and glycine solutions were selected, whereas in protein sufficiency umami substance solutions were chosen (Fig. 13).

Spontaneously hypertensive rats (SHR) have the tendency to ingest a large amount of sodium chloride and become hypertensive; however, sodium intake declined as the protein content in the diet was increased. When an umami substance was available as an aqueous solution, rats selected it, and a drastic decrease in total sodium intake—even up to 70%—was observed when the solutions were offered in conjunction with the diet of higher protein content. Although the predisposition of SHR to hypertension was essentially unchanged, the pathogenesis of atheroma was actually depressed when the diet was sufficient in protein and umami substances were available (Figs 14a and 14b).

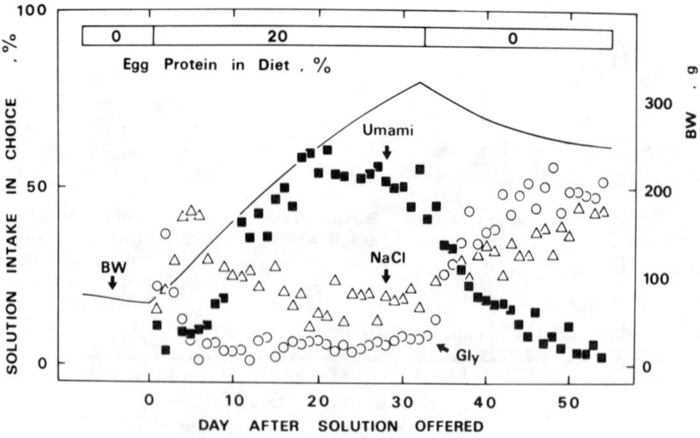

FIG. 13. Dietary protein content and preference.

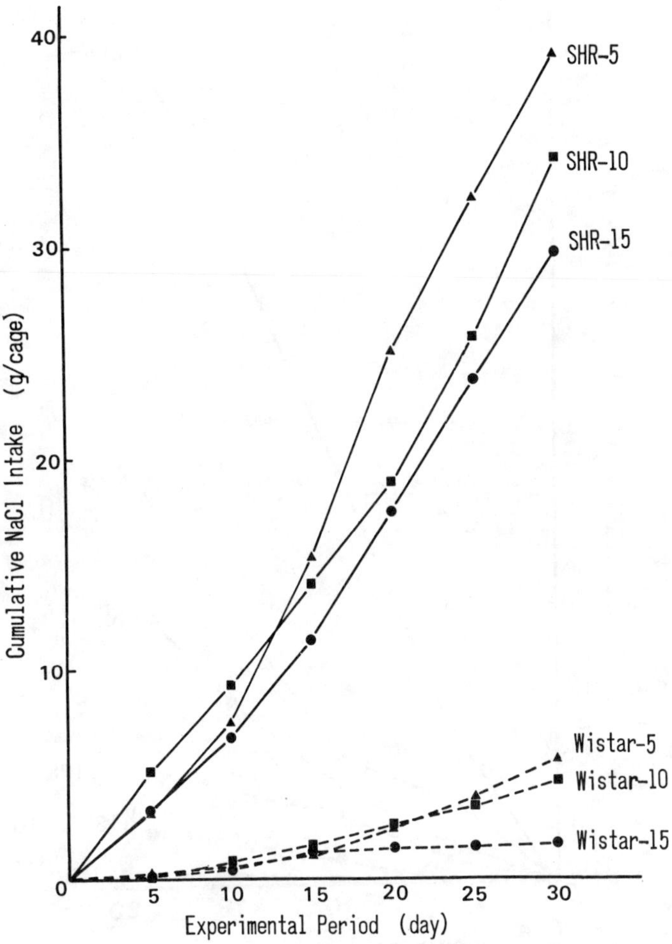

FIG. 14a. Changes in cumulative sodium chloride (NaCl) intake of SHR, Wistar rats and SD rats fed 5%, 10% and 15% PEP diet.

Rats display an ability to select a balanced amino acid diet over an imbalanced one. This choice is maintained in quinine-adapted rats even when quinine is added to the balanced amino acid diet and the much-preferred saccharin is added to the imbalanced diet (Fig. 15), indicating that flavour preferences can be overriden by metabolic needs. When an L-lysine-deficient diet is offered, rats discriminate an L-lysine solution among 15 kinds of L-amino acid

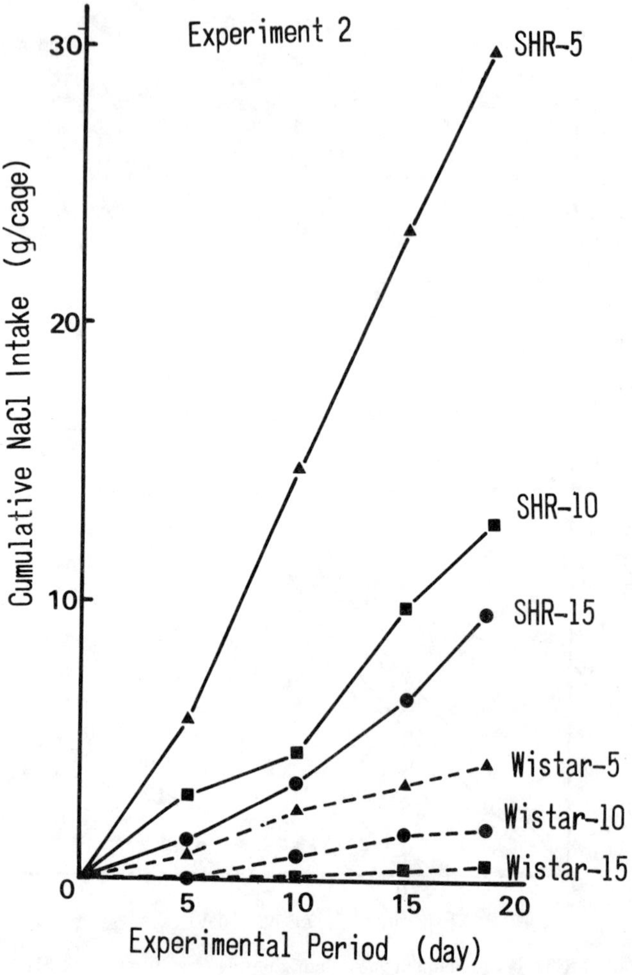

FIG. 14b. Changes in cumulative sodium chloride (NaCl) intake of SHR, Wistar rats and SD rats fed 5%, 10% and 15% PEP diet (+IMP, MSG).

solutions and ingest a sufficient quantity of solution to supplement the L-lysine-deficient diet. Once the L-lysine deficiency has been alleviated, rats select solutions of umami substances. These results suggest that taste preference is dependent on the quality as well as the quantity of protein in the diet. Thus preference for umami may well be a taste marker for adequate protein nutrition.

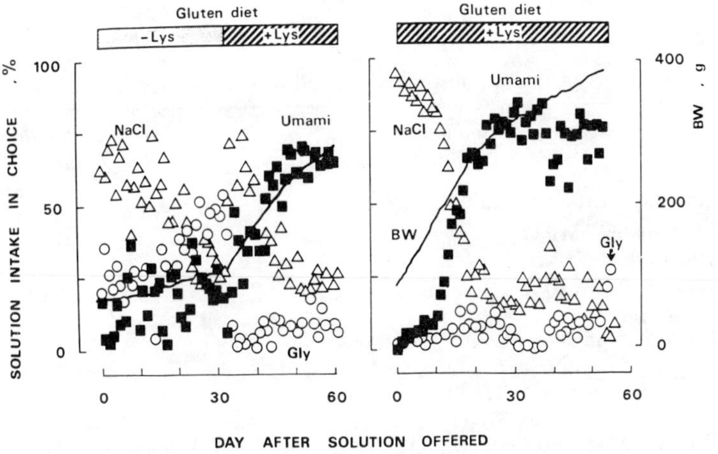

FIG. 15. The relations between umami and the balance of protein quality and quantity.

CONCLUSION

The classification of taste qualities into the four basic tastes was advanced before modern taste physiology had developed, yet has been applied *a priori* even up to the present time. Studies on receptor mechanisms of umami substances in animals have shown that receptor sites for umami are independent of those specific to the four basic taste stimuli. Single-fibre analysis of units in the mouse glossopharyngeal nerve also suggests the existence of units sensitive only to umami. Likewise, multidimensional analysis of psychophysical data in humans indicates that umami is independent of any of the four basic tastes. Thus results obtained by both neuroelectrophysiological and psychophysical studies favour the conclusion that umami is a taste sensation independent of the four basic tastes. Psychophysical and electrophysiological experiments have also demonstrated that Glu itself has no enhancing effect on the intensities of any of the four basic tastes, although there is a remarkable synergism between Glu and nucleotides (Table 1). The flavour potentiation induced by the addition of Glu to foods seems to be due to its synergistic effect with nucleotides contained in foods. Further study would be needed to clarify the mechanism of this flavour potentiation which is unique to umami taste.

TABLE 1
Umami Substances Related to Nucleotides

Substance (disodium salt)	*Relative potency of umami*
5′-Inosinate, 7·5H_2O	1
5′-Guanylate, 7H_2O	2·3
5′-Xanthylate, 3H_2O	0·61
5′-Adenylate	0·18
Deoxy 5′-guanylate, 3H_2O	0·62
2-Methyl-5′-inosinate, 6H_2O	2·3
2-Ethyl-5′-inosinate, 1·5H_2O	2·3
2-Phenyl-5′-inosinate, 3H_2O	3·6
2-Methylthio-5′-inosinate, 6H_2O	8·0
2-Ethylthio-5′-inosinate, 2H_2O	7·5
2-Ethoxyethylthio-5′-inosinate	13
2-Ethoxycarbonylethylthio-5′-inosinate	12
2-Furfurylthio-5′-inosinate, H_2O	17
2-Tetrahydrofurfurylthio-5′-inosinate, H_2O	8
2-Isopentenylthio-5′-inosinate (Ca)	11
2-(β-Methallyl)thio-5′-inosinate	10
2-(γ-Methallyl)thio-5′-inosinate	11
2-Methoxy-5′-inosinate, H_2O	4·2 (3·7)
2-Ethoxy-5′-inosinate	4·9
2-*i*-Propoxy-5′-inosinate	4·5
2-*n*-Propoxy-5′-inosinate	2
2-Allyloxy-5′-inosinate (Ca), 0·5H_2O	6·5
2-Chloro-5′-inosinate, 1·5H_2O	3·1
N^2-Methyl-5′-guanylate, 5·5H_2O	2·3
N^2,N^2-Dimethyl-5′-guanylate, 2·5H_2O	
N^1-Methyl-5′-inosinate, H_2O	0·74
N^1-Methyl-5′-guanylate, H_2O	1·3
N^1-Methyl-2-methylthio-5′-inosinate	8·4
6-Chloropurine riboside 5′-phosphate, H_2O	2
6-Mercaptopurine riboside 5′-phosphate, 6H_2O	3·4
2-Methyl-6-mercaptopurine riboside 5′-phosphate, H_2O	8
2-Methylthio-6-mercaptopurine riboside 5′-phosphate, 2·5H_2O	7·9
2′,3′-*O*-Isopropylidene 5′-inosinate	0·21
2′,3′-*O*-Isopropylidene 5′-guanylate	0·35

BIBLIOGRAPHY

1. Kawamura, Y. and Kare, M. R. (1986) *Umami—A Basic Taste—Physiology, Biochemistry, Nutrition, Food Science*. Marcel Dekker Inc, New York.
2. Kawamura, Y., Kurihara, K., Kato, H., Konosu, S., Kimura, S. and Yamaguchi, S. (1985) *Rinsho Eiyo* [The Japanese Journal of Clinical Nutrition] **66**, 129–65.
3. Sugita, Y. (1985) Umami: Its Physiology and Psychology. *Shokuhin to Kagaku* [Food and Science] **27**, 103–6.
4. Sugita, Y. (1986). Umami: Its Biological meaning. *Information on Dietary Life*. No. 10, ed. Ajinomoto Co., Inc, Tokyo.
5. Umami Information Center (1984) *Sweet, Sour, Salty, Bitter and Umami.* Tokyo.

5

The Use of Computers in Flavour Development

WILLIAM I. TAYLOR
International Flavors and Fragrances (EAME), London, UK

ABSTRACT

Progress in any field is usually the result of much hard work notwithstanding the fact that the target is well conceived. The collection and recording of data is often the most time-consuming part of the effort. In the past the software may have been a deck of filing cards and the hardware pencil and paper. Because of the phenomenal advances made in computer hardware and software in recent years, advantage can be taken of the low cost and speed in processing to apply this technology to any field. Major applications to flavour research and development lie in flavour analysis, handling of structure–activity relationships and data-use. Examples are given.

INTRODUCTION

Why should a discussion of the use of computers in flavour development be included in this symposium? One reason could be because there are clearly understood advantages for their use peculiar to flavour research and development. Another reason could be that somehow computers make possible conclusions not accessible by other techniques. The answer is that one will always use the tools which enable a job to be carried out more efficiently. If the use of a pencil and a few filing cards will accomplish the task in the same time there is no need for a computer. The advantage lies with the computer because of accurate data storage and high-speed processing. However, all this would be academic but for the availability of low-cost computers (the 'hardware') and user-friendly processing methods (the 'software'). This statement is good for hardware of any size and especially true for software for micro- and mini-computers. Where large quantities of data must be handled, a properly programmed computer can do in seconds what in the past could have taken days, weeks or even years to complete. The other advantage of the computer is that once it has been given a set of selection rules, that is 'programmed', it will tirelessly labour until the entire file has been processed. Apart from speed a real plus is that the data once loaded are always available for re-examination in many ways, even including data the user may have forgotten about. Nevertheless, the computer works precisely, and only, within the framework of the programmed instructions. It does not replace thinking carefully about the proper use of purchased software.

FLAVOUR ANALYTICAL SUPPORT

If foodstuffs were taken directly for use from nature without storage, processing or cooking, there would have been little need for flavour research, apart from academic curiosity. With the advent of food preservation and convenience foods, there came the added-value possibilities of putting lost flavour back into these products. Thus came the need for flavour research and development. If ever an analytical technique was invented which was tailor-made for this research it was the invention of gas–liquid chromatography. It enabled, for the first time, the accurate and quantitative assay of micro quantities of volatile mixtures (for some results see Table 1). Over the last thirty years its value has increased in proportion to the way microprocessors have been developed to automate the

TABLE 1
Amount of Volatiles in Selected Foods

Food	*Volatiles (ppm)*	*Components*
Beef	50	>300
Coffee	1	~500
Passion fruit	12	~200
Strawberry	10	~300
Tea	60	~300

TABLE 2
Typical Cost for a GC–MS with Data Processing

Part	*Function*	*Cost (£)*
GC–split inlet for 2 capillary interface	Molecule separation	~9 000
Autosampler and injector	Unattended operation	~4 000
Mass detector with microprocessor, data storage and printer	Structure analysis	~50 000
Application software and library data	Data massaging and searching	~6 000
	Total	~70 000

assays, and further use of the mass spectrometer, to analyse the eluting separated components and reduce data collection to a problem of data analysis. With further data processing capability, it is now possible to inject a micro sample into the gas chromatograph and receive at the end of the run a report listing the percentage of each component and the identification of the constituents known to the data base (see Table 2). Today compared with twenty years ago the rewards come not from generating the data but from what one does with it. Everyone can cite cases where collecting data used to be the larger part of a day's work. The amount of time spent in just collecting information is decreasing dramatically and will continue to do so. I remember the time when state of the art precise weighings used to be carried out using a double-beam balance and when an infrared spectrum only made sense after the ambient background was laboriously subtracted from the output of a single-beam machine. The analytical equipment used today depends integrally upon microprocessor control to the benefit of flavour research.

While knowledge in flavour chemistry has in the past advanced at a rate inversely proportional to the time taken to carry out analyses, this rate of discovery is now slowing down, at least for individual volatile flavour molecules, because the bulk of them are now known. Nevertheless exciting discoveries are coming from the detection and characterisation of trace ingredients, among others, pyranones, furanones, thiophenes, thiazoles, oxazoles, oxazolines, pyrroles, pyrazines, many of which are only formed during the

TABLE 3
Sulphur Compounds of High Potency

Name	*Source*
D-*trans*-8-Mercapto-*p*-menthan-3-one	Blackcurrant principle in Buchu leaves[1]
Propanethial-*S*-oxide	Lachrymatory principle of onions[2]
3,5-Dimethyl-1,2,4-trithiolane	Boiled beef[3]
2-Methyl-4-propyl-1,3-oxathiane	Passion fruit[4]
1-*p*-Menthene-8-thiol	Grapefruit[5]

boiling or roasting of food. Often the most difficult to isolate and characterise are the sulphur compounds (see Table 3) because they occur at levels which test both the analytical equipment and the flavourist. Successful detection has depended on the close collaboration of the GC–MS specialist (stretching the analytical capability to the limit, $\sim 10^{-12}$ moles) and the flavourist's nose. The flavourist has the problem of recognising the hedonic flavour value of a component being smelled in the gas eluting from the gas chromatographic column. Character impact value is often difficult to assess by itself and at an odd concentration. Without computer-assisted processing, useful results of this sort would be hard to get. Given a complete knowledge of all the volatile flavour molecules, there still remains an enormous amount to be learnt as to how combinations of volatile ingredients donate useful organoleptic values. There is also virgin territory in understanding the role that non-volatile ingredients play in flavour appreciation, the role of temperature and mouthfeel effects and supporting inputs from the senses of sight, sound and touch.

MOLECULAR MODELLING

Many and continuing efforts are being made to rationalise the connection between the structures of organic compounds (including molecules detectable by the chemical senses) and their biological activities. This comprises the field of structure–activity studies. Correlations between molecular structure and biological effect are important for the development of pharmacological agents, herbicides, pesticides, and the investigation of chemical and genetic toxicity. Practical importance attaches to these studies because the results can be used to predict the activity of untested compounds. In addition, the technique can direct a researcher's attention to molecular features that correlate highly with biological activity. Structure–activity relations have been applied largely to research in the pharmaceutical and agricultural industries.

Obviously the better way to develop predictive capability is to understand, at the molecular level, the mechanisms that lead to the biological activity of interest. Unfortunately this knowledge is not available, except in rare cases, for substances which react with the chemical senses. This is because the researcher has to depend on poorly articulated descriptions of emotional responses to flavour stimuli. At best descriptions are qualitative and much dependent on prior experience. Articulation is the last step in a chain of events which starts with the transport of the odorant from the air stream to the nasal receptor or receptors, followed by transduction of the signal or signals to the olfactory lobe for further processing along with other sensory inputs. This can complicate the interpretation of the correlations which result from a structure–activity analysis. Approaches to structure–activity relationships which have been used include the semi-empirical linear free energy approach proposed by Hansch,[6] the additivity or free-Wilson model,[7] quantum mechanically based models,[8] multidimensional scaling techniques[9] and pattern recognition methods.[10]

I take it for granted that the molecular shape and olfactory type are directly related. I rest my case on the evidence throughout molecular biology that molecules fit onto specific sites in order either to be transformed or to initiate transmission of a signal. The assumption made in the pattern recognition method is that the structures of compounds having a particular odour sensation, and similar structural classes that do not, can be adequately represented by a set of molecular descriptors. A relation can be discovered between the structure and the activity by applying statistical and pattern recognition techniques to a set of tested compounds. The relation can be extrapolated to untested compounds. The problem in practice is to find a set of adequate descriptors for the set under consideration, a task only made possible by the use of suitable computer software. The fundamental steps are:

(1) Identify, assemble, input, store and describe a data set of structures for the chemicals.

(2) Develop computer generated molecular descriptors for members of the set. These descriptors may be derived directly from the stored topological representations of the structures or they may require the development of three-dimensional molecular models.

(3) Test the predictive ability of these discriminants on compounds of unknown activity.

(4) Systematically reduce the set of molecular structure descriptors employed to the minimum set sufficient to retain discrimination between active and inactive compounds. No matter what analytical technique is used to determine the structure–activity relationships for volatile flavour odorants, the results have not been clear-cut. Probably this is because the descriptors cannot be allotted to the steps referred to above, namely transport to the detector, reactivity with the receptor(s), and interaction with other signals, all of which must happen before the input can be described in words.

A study of musks has been carried out by the pattern recognition technique, using a data set made up of 140 musks along with 70 chemicals selected at random from a group of 109 non-musks. The investigators required 13 descriptors to define musk character. The program was able to classify, correctly, 183 out of the 210 chemicals, with a success rate of 133 out of the 140 musks and 50 out of 70 of the non-musks.[11] The conclusion is that, while progress has been made, at this stage the accurate prediction of musk character by the examination of the chemical structure alone is not possible.

Despite the seductive processing power of computers, progress has already been made in taste and smell by the application of common sense. One is for sweetness. The Shallenberger and Acree AH-B rule[12] has proven to be a very useful generalisation. The other recognises the importance, for ambergris odours, of a 1,2,4-triaxial arrangement of substituents (one of them an oxygen) on a decalin ring.[13]

SELECTION OF FLAVOURS

As a final example of the use of computers in flavour development, allow me to comment briefly on the use of computers for the storage and retrieval, on a selective basis, of flavour formulae for commercial application. Parenthetically, a selling flavour is really the only good flavour! For a given flavour creation, the master record is keyed to a reference number which uniquely identifies the formula. The file contents must include all the elements needed to enable rapid sorting to provide a flavour or a group of flavours which meet the selection parameters. These data elements (see Table 4) include code number, price, end-use categories, flavour descriptors, flavour appearance and colour, form of use, solubility, solvents and carriers, additives with E and T numbers, special ingredients, storage conditions, and legal status. In other words, the record contains data concerned with flavour performance as well as information to enable the legal status for the use and the country of destination to be programmed into the selection process.

It is inevitable that a file of this sort could contain formulae with respectively differing descriptors, but having identical performance properties for a given end-use condition. It serves no useful purpose to be able to discriminate between two flavours in a side-by-side comparison if the consumer cannot distinguish them when they are consumed apart. This problem is controlled to a large extent by allowing only approved words to be used for hedonic descriptions and by further qualification using brand names wherever relevant. To prevent the file from having too many creations of the same

TABLE 4
Flavour Selection Parameters

Field	*Examples*
Reference number	XYZ
Flavour category	Beverage, dairy
Selling—type	Best, restricted
—form	Emulsion, powder
—related forms	GRAS version, LPV
Descriptor—primary	Apple, cereal
—secondary	Conc., baked
—brand name	Delmonte, CPC
Solubility	Water, oil/fat
Shelf life	<1 mth, 3 mths
Storage conditions	Deep frozen, RT
Product test results	Satisfactory, N/A
Legal—aromatic part	AA, AI, NI, NN
—whole flavour	WONF, Parve
—solvents	Acetic acid, DPG
—carriers	Gum arabic, salt
—additives	E and T numbers
—special ingredients	Vanillin, maltol
Special comments	As required

type and for the same end-uses, the file can be arbitrarily frozen at an agreed-upon size. This requires a single file administrator, who enforces also the selection rules for product entry and removal. Among other options, the file should be of a defined size and, once that size is reached, a flavour has to be removed before a new one can be entered, and this irrespective of the application. To be entered, a flavour must have test data to prove that it performs in the specified end-use and satisfies legal and stability requirements. Elimination of a flavour formula has to be based on rules which the computer can execute. Elimination can be made automatic, following rules as simple or as refined as is wished. For example, a flavour could be removed because it has not been selected or sold two years after entry in the system. It is obvious that, with good reason, a given flavour could be reinstated.

REFERENCES

1. LAMPARSKY, D. and SCHUDEL, P. (1976). *Tetrahedron Letters*, 3323.
2. BRODNITZ, M. H. and PASCALE, J. V. (1971). *J. Agric. Food Chem.*, **19**, 269.
3. WINTER, M., FURRER, A., WILLHALM, B. and THOMMEN, W. (1976). *Helv. Chim. Acta*, **59**, 1613.
4. CHANG, S. S., HIRAI, C., REDDY, B. R., HERZ, K. O., KATO, A. and SIPMA, G. (1968). *Chem. and Ind.*, 1639.
5. DEMOLE, E., ENGGIST, P. and OHLOFF, G. (1982). *Helv. Chim. Acta*, **65**, 1785.
6. HANSCH, C. (1979). In: *Advances in Linear Free Energy Relations*, Vol. 2, Eds Chapman, N. R. and Shorter, J., Plenum Press, New York.
7. CRAIG, P. W. (1972). In: *Biological Correlations: The Hansch Approach*, Ed. van Walkenburg, W., Amer. Chem. Soc., Washington, DC.
8. CHRISTOFFERSEN, R. E. (1976). In: *Quantum Mechanics of Molecular Conformations*, Ed. Pullman, B., Wiley, New York.
9. SCHIFFMAN, S. S., REYNOLDS, M. L. and YOUNG, F. W. (1981). *Introduction to Multidimensional Scaling: Theory, Methods and Applications*, Academic Press, New York.
10. STUPER, A. J., BRUGGER, W. E. and JURS, P. C. (1979). In: *Computer Assisted Studies of Chemical Structure and Biological Function*, Wiley-Interscience, New York.
11. JURS, P. C., HAM, C. L. and BRUGGER, W. E. (1981). In: *Odor Quality and Chemical Structures*, Eds Moskowitz, H. R. and Warren, C. B., Amer. Chem. Soc. Symposium Series 148, Washington, DC.
12. SHALLENBERGER, R. S. and ACREE, T. E. (1967). *Nature*, **216**, 480.
13. OHLOFF, G., GIERSCH, W., SCHULTE-ELTE, K. H. and VIAL, C. (1976). *Helv. Chim. Acta*, **59**, 1140.

6

Biogeneration of Plant Aromas

P. SCHREIER

Lehrstuhl für Lebensmittelchemie, Universität Würzburg, Federal Republic of Germany

ABSTRACT

After a general introduction into the biogeneration of plant aromas displaying different pathways of secondary plant metabolism special attention is focused on the biotransformation of monoterpene alcohols and derivatives. Thus the possible role of diols as flavour precursors is discussed. Furthermore, particular emphasis is devoted to epoxidation reactions and their subsequent transformations. Finally, glycosylation as an essential step of monoterpene catabolism is discussed.

INTRODUCTION

From ancient times plant volatiles have been of considerable significance in relation to flavour and perfumery.[1,2] Today they continue to play an important role as natural odorous compounds and to serve as models for chemical synthesis of 'nature-identical' substances. In general, plant volatiles, including essential oils, comprise a high number of constituents occurring in complex compositions and in distinct concentration ranges in various plant tissues. They are formed from non-volatile precursors in two types of enzymic reactions:[3] intracellular pathways, and reactions occurring due to disruption of cell tissue.

BIOGENERATION PATHWAYS DUE TO DISRUPTION OF TISSUE

In most vegetables, thio compounds and carbonyls are the predominant volatiles. The intact tissue contains only non-volatile precursors, well separated from the appropriate enzymes. During disruption of tissue, e.g. by homogenisation or cutting, the non-volatile precursors are transformed into volatiles, in particular by the action of enzymes and, to a smaller extent, by heating during processing. The fundamental reactions leading to volatiles from non-volatile precursors due to disruption of tissue have been summarised by Tressl *et al.*[4] The precursors comprise *S*-alkyl- and *S*-alkenyl-cysteine sulphoxides, glucosinolates and unsaturated fatty acids.

Volatiles from *S*-Alkyl- and *S*-Alkenyl-cysteine Sulphoxides

S-Alkyl- and *S*-alkenyl-cysteine sulphoxides are the typical non-volatile aroma precursors of *Allium* genus, which comprises important edible vegetables such as e.g. onion (*A. cepa* L.), garlic (*A. sativum* L.), leek (*A. ampeloprasum* L.) and shallot (*A. cepa* var. *aggregatum* L.). C-S-Lyases (allinases) are responsible for the enzymic splitting of non-volatile precursors (Fig. 1). The enzymes, the precursors and the products have been extensively studied.[5,6] Recently new results about the chemistry of garlic and onion have been published.[7,8] As outlined in Fig. 2, due to cell decompartmentation, e.g. in garlic, allicin, formed by dimerisation of intermediate sulphenic acid (Fig. 1), is decomposed to a series of various secondary products. Among them the newly identified, physiologically active isomeric ajoenes as well as the vinyl-dithiines have to be stressed (Fig. 2).

Volatiles from Glucosinolates

Glucosinolates, substances related to the widely distributed cyanoogenic glycosides, display the general formula outlined in Fig. 3. Enzymic degradation by myrosinase (thioglucoside glucohydrolase, EC 3.2.3.1) activity, after tissue rupture, leads to D-glucose, sulphate, isothiocyanates, organic cyanides and sulphur, or thiocyanates. Almost all plants containing glucosinolates belong to the family Cruciferae, which comprises several *Brassica* species, such as e.g.

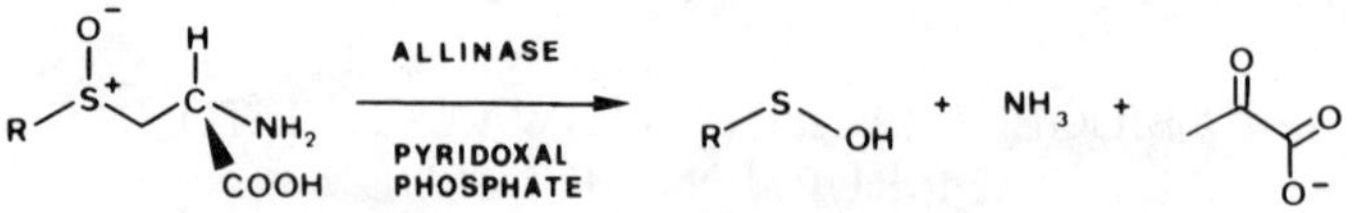

FIG. 1. Formation of sulphenic acid, RSOH, along with ammonia and pyruvate from non-volatile precursor by allinase activity (e.g. if R is an allyl group, C_3H_5, the precursor is allicin).

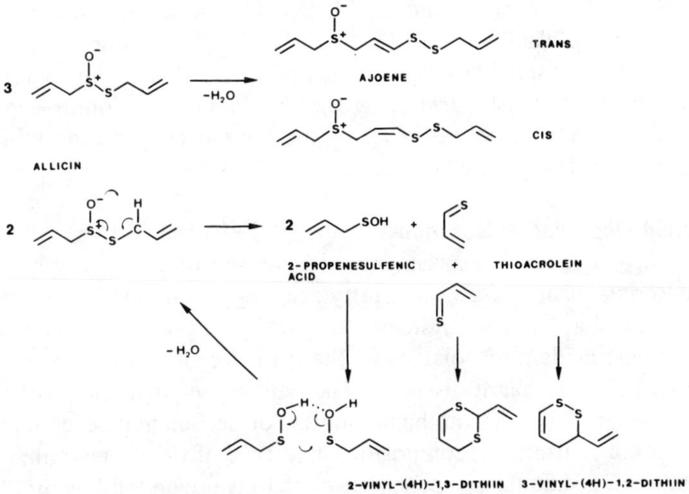

FIG. 2. Decomposition of allicin, formed by dimerisation of RSOH (cf. Fig. 1), in several pathways.[7,8] Top: formation of two molecules of ajoene from three molecules of allicin. Bottom: decomposition of allicin to 2-propenesulphenic acid and thioacrolein. The self-condensation of two molecules of 2-propenesulphenic acid regenerates allicin, and vinyl-dithiines are produced from thioacrolein by Diels–Alder reaction.

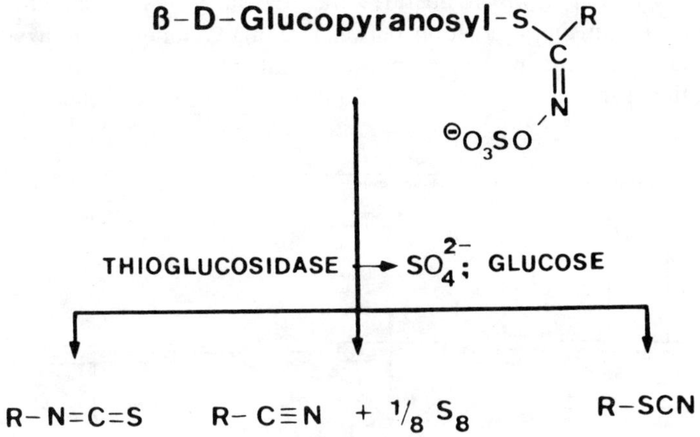

FIG. 3. Enzymic degradation of glucosinolates yielding D-glucose, sulphate, isothiocyanates, organic cyanides, sulphur and thiocyanates.

cabbage (*B. oleracea capitata* L.), cauliflower (*B. oleracea* var. *botrytis* L.), black mustard (*B. nigra* L.), brown mustard (*B. juncea* L.), white mustard (*B. alba* L.) as well as radish (*Raphanus sativus* L.) and horseradish (*Armoracia lapathifolia* Gillb.). Comprehensive reviews on glucosinolates covering specific aspects, e.g. biosynthesis, are available.[9]

Lipid Degradation Reactions

Early incubation experiments with homogenates of various plant materials, using unsaturated fatty acids as substrates, have demonstrated that enzyme systems specifically degrading linoleic and linolenic acids into volatiles in the presence of oxygen are widely distributed in plant tissues.[3] The sensory contribution of such 'secondary' volatiles can be favourable or detrimental depending on their concentrations, composition and type of plant material. Thus they can accumulate in concentrations to cause an 'off-flavour', but, on the other hand, there are examples, e.g. cucumber or melon, where the typical aroma is characterised by these 'secondary' volatiles.

The key step in the formation of 'secondary' volatiles derived from linoleic and linolenic acids is the hydroperoxidation of these acids by lipoxygenase (EC 1.13.11.12), an enzyme displaying high substrate specificity for the *Z*,*Z*-1,4-pentadiene structure.[10] This enzymic oxidative degradation of lipids is preceded by the action of acyl hydrolase, which liberates the free fatty acids. Galliard's[11] remarks on the interaction between cellular damage and enzymic lipid destroying pathways in potato tuber can be generalised as outlined in Fig. 4.

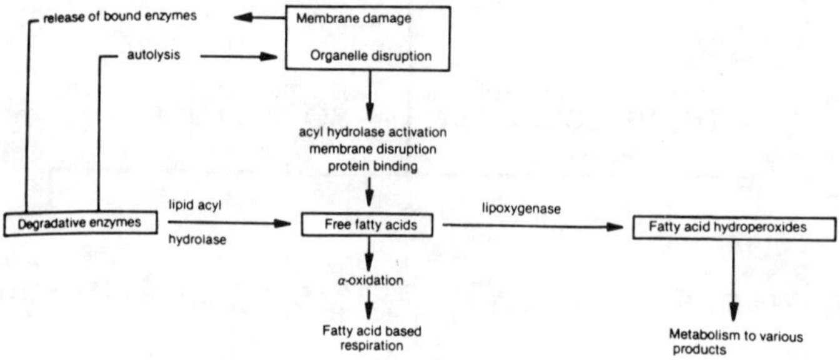

FIG. 4. Interactions between cellular damage and lipid degrading enzymes.[11]

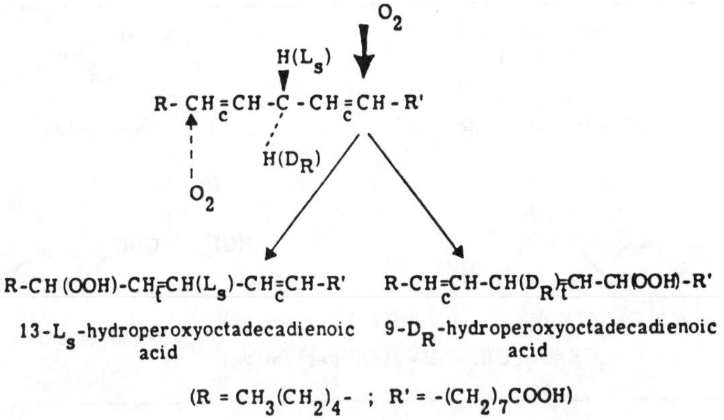

FIG. 5. Stereospecific formation of isomeric hydroperoxides from linoleic acid by lipoxygenases.[11]

Based on the previous work of Hamberg and Samuelson[12] and Egmond *et al.*,[13] Galliard[11] summarised the stereospecific formation of isomeric hydroperoxides from linoleic acid by lipoxygenases as shown in Fig. 5. The initial step is the stereospecific abstraction of a hydrogen atom from the methylene group of the pentadiene system; electronic rearrangement and stereospecific attack by molecular oxygen result in an optically active conjugated hydroperoxy fatty acid. The classical soybean enzyme (type 1 isoenzyme) produces preferably 13-L-hydroperoxide from linoleic acid, whereas e.g. that from corn germs forms the D-9-isomer. The information available on the peroxidation specificity of lipoxygenases from different sources has been reviewed by Grosch.[14]

The enzymic cleavage of fatty acid hydroperoxides to carbonyl fragments is schematically outlined in Fig. 6.[11] Thus the volatile products from the 9-hydroperoxides of linoleic and linolenic acids are *Z*-3-nonenal and *Z*,*Z*-3,6-nonadienal, respectively, which represent two main volatile components of the characteristic cucumber odour.[15] Enzymic cleavage of 13-hydroperoxide leads to the formation of hexanal and *Z*-3-hexenal.[11] The labile membrane-bound hydroperoxide lyase has been partially purified from various fruits[11,16,17] and, in particular, from tea leaves.[18] As recently demonstrated by our group,[19] in contrast to fruits, the enzyme from leaves is bound, probably as an extrinsic enzyme, to the thylakoid membrane of chloroplasts. The unsaturated alkenals originating

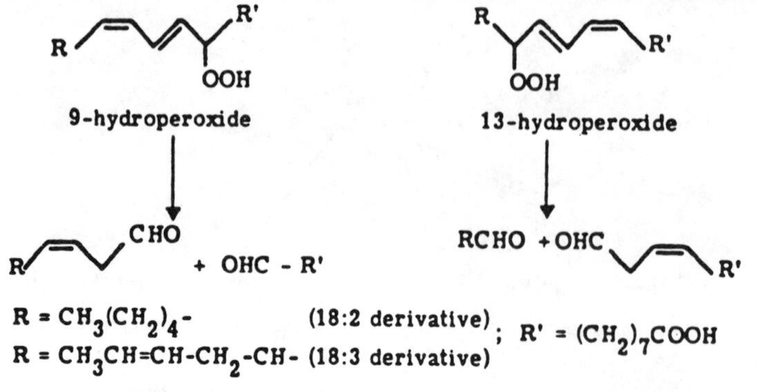

FIG. 6. Enzymic cleavage of fatty acid hydroperoxides to carbonyls by hydroperoxide lyase activity.[11]

from the lipoxygenase-hydroperoxide cleaving enzyme can be further transformed by ADH activity. They are not only reduced to the alcohols, but are also partly metabolised into the corresponding saturated aldehydes. Thus, in the presence of NADH, incubation of e.g. *E*-2-hexenal with tomato ADH led to formation of hexanal, hexanol, and *E*-2-hexenol.[20]

As outlined in Fig. 7, different amounts of C_6 volatiles have been determined in leaves from various plants.[19] With concentrations

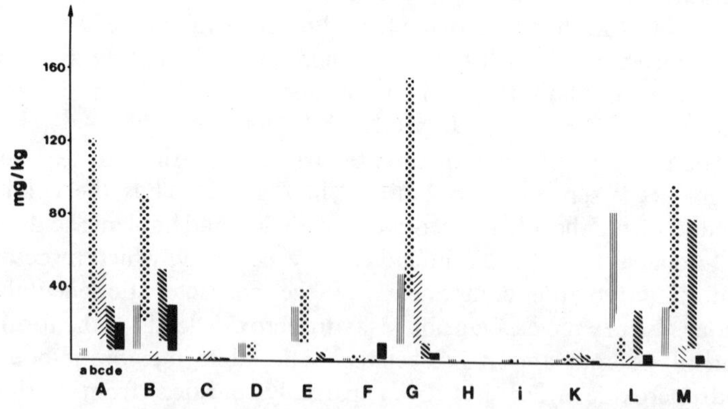

FIG. 7. Amounts of C_6 volatiles (ranges, mg/kg fr. wt) determined in leaf homogenates from various plants.[19] a = hexanal; b = *E*-2-hexenal; c = 1-hexanol; d = *Z*-3-hexen-1-ol; e = (*E*)-2-hexen-1-ol. A = vine; B = tobacco; C = spinach; D = chicory; E = leek; F = lettuce; G = endive; H = cabbage; I = savoy; K = feeding turnip; L = kohlrabi; M = radish.

TABLE 1
Summary of Known Enzymic Transformations of Unsaturated C_{18} Fatty Acid Hydroperoxides in Plant Extracts[11]

Reactions	*Enzyme*	*Major products*
Reduction	'Lipoperoxidase'	C_{18} hydroxy acids
Isomerisation	Hydroperoxide isomerase	C_{18} α- and β-ketol acids
Isomerisation/ dehydration	Enzyme in potato tuber	C_{18} divinyl ether acids
Chain cleavage	Cleavage enzyme (hydroperoxide lyase)	C_6 or C_9 aldehydes C_{12} or C_9 oxo acids
Isomerisation/ hydration	?	C_{18} epoxy-hydroxy acids C_{18} trihydroxy acids
Anaerobic reactions	Lipoxygenase	C_{18} keto acids C_{18} keto acids + pentane

each of >50 mg/kg, leaf homogenates from vine (A), tobacco (B), endive (G), kohlrabi (L) and radish (M) have been found to be highly active in C_6 aldehyde formation.

Since fatty acid hydroperoxides are cytotoxic and affect protein and membrane structures, there are many pathways in plants to metabolise these compounds. Galliard[11] has summarised the various enzymic transformation pathways of fatty acid hydroperoxides in plants (Table 1).

Summarising, it can be concluded that, for the biogenetic pathways of 'secondary' plant volatiles, extensive but still incomplete information is available. In particular, the control mechanisms of biogenesis and accumulation still await elucidation.

INTRACELLULAR BIOGENETIC PATHWAYS

The biogenetic pathways of natural plant volatiles can be derived from the enzymically controlled lipid, terpene, amino acid, and phenylpropane metabolism of the plant.[4] Certainly, the principal pathways are well known and constitute part of several textbooks. But if one asks for details often an answer will be lacking. Thus, for the biogeneration of lipid metabolites such as aliphatic esters, alcohols, acids, carbonyls and lactones, a plausible concept has been developed,[21] but the concrete question, in which manner e.g. are γ-lactones biogenerated, can only be hypothetically answered as

(i) reduction of 4-oxoacids, (ii) hydration of unsaturated acids, (iii) chain elongation of 3-hydroxy-acids, or (iv) hydroxylation of aliphatic acids. As to hydroxy-γ-lactones, the hydrolysis of Z-α,β-unsaturated γ-epoxy-esters has also to be considered.[22] For all these proposals about the metabolism of plants, there is no experimental evidence as yet!

A similar situation will be found if one discusses the terpenes; among them monoterpenes constitute an important class of plant volatiles. A recently published encyclopaedic handbook on the chemistry of monoterpenes[23] comprises approximately 1500 text pages; less than 5% deal with the biogenesis of these compounds. On the other hand, our extensive knowledge of chemical structures and non-enzymic reactions of monoterpenes is evident. This situation enables us to develop a strategy to extend our still-limited information about the biogeneration of monoterpenes. First of all, we should deepen our knowledge on monoterpene structures by *identifying natural monoterpenes (and derivatives) whose structures are known from non-enzymic reactions*. Based on this concept, secondly, it will be necessary to increase activities in *performing enzymic studies of formation and catabolism of monoterpenes*.

Despite the great physiological importance of the catabolism of monoterpenes, practically nothing is known about the pathways or mechanisms of such catabolic steps. At present, we only know from many labelled time-course studies combined with investigations of diurnal variation and longer-term periodic analyses that monoterpenes are not stable end-products of secondary metabolism. They can undergo biochemical turnover and can even be recycled into primary metabolism.[24] Such metabolic reactions imply the transformation of the lipophilic monoterpenes into more hydrophilic structures. As to such hydrophilisation, introduction of hydroxyl functions yielding 'polyols' and/or glycosides[24,25] has been discussed as an essential step in monoterpene metabolism.

In the following, some results from our recent work on monoterpenes from tropical fruits will be presented, in which we followed the above-mentioned strategy, i.e. first of all to look for structures still unknown as natural plant constituents.

Monoterpene 'Polyols'

Using a sample preparation technique suitable for the isolation and enrichment of polar compounds from plant tissue,[26] we investigated some oxygenated monoterpenes derived from linalool.[27] This monoterpene alcohol is, together with benzyl isothiocyanate, the major volatile constituent in papaya (*Carica papaya* L.) fruit pulp.[28,29] As hydroxylated linalool derivatives, identified in papaya fruit by capillary gas chromatography (HRGC) and coupled capillary gas chromatography–mass spectrometry (HRGC–MS), the compounds **1–5** were found (Fig. 8). The monoterpene 'polyols' **1–3** had already been detected in cinnamon,[30] grapes[26,31,32] and passion fruit,[33] but not in papaya.

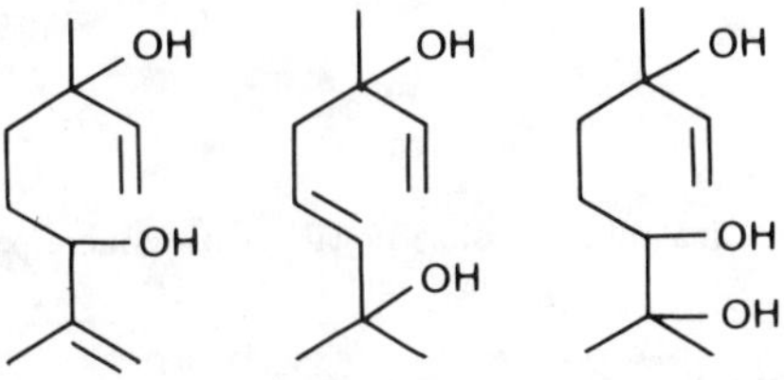

1 **2** **3**

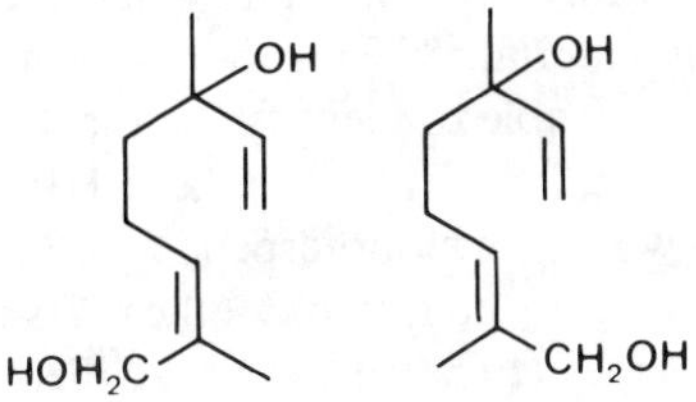

4 **5**

FIG. 8. Structures of hydroxylated linalool derivatives identified in papaya (*C. papaya* L.) fruit pulp by HRGC and HRGC–MS.[26] **1** = 2,6-dimethylocta-1,7-diene-3,6-diol; **2** = 2,6-dimethylocta-3,7-diene-2,6-diol; **3** = 2,6-dimethyloct-7-ene-2,3,6-triol; **4**,**5** = *Z*- and *E*-2,6-dimethylocta-2,7-diene-1,6-diol, respectively.

The isomeric diols **4** and **5** were determined in a ratio of about 1:6 (*Z*:*E*) in the fruit pulp. Biogenetic formation via enzymic ω-hydroxylation seems plausible as previously proposed for several bacterial[34,35] and fungal[36] transformations of linalool. In higher plants, the *E*-isomer **5** has been found in different *Nicotiana* species.[37,38] The β-D-glucosides of *E*-**5** and *Z*-2,6-dimethylocta-2,7-diene-1,6-diol (**4**) have been isolated from *Betula alba* leaves and the fruits of *Chaenomeles japonica*.[39]

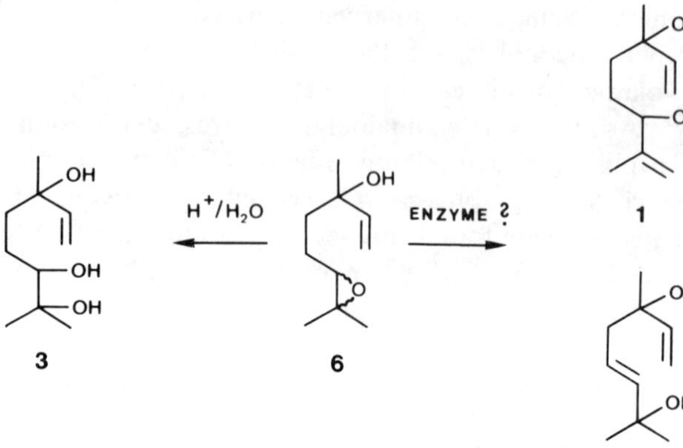

2

FIG. 9. 6,7-Epoxy-linalool as possible precursor of monoterpene diols **1** and **2** as well as triol **3**.

6,7-Epoxy-linalool

Previously, the isomers of 6,7-epoxy-linalool (**6**) had been proposed as the potential precursors of the monoterpene 'polyols' **1–3** (Fig. 9), but without any experimental evidence for the occurrence of **6** in natural plant material.[26] Using the above-mentioned isolation technique we were now able to identify the diastereoisomeric epoxy derivatives **6** in papaya fruit by HRGC and HRGC–MS.

During previous work, the monoterpene triol **3** had been suggested as a possible precursor of the hydroxy-ethers **7** and **8**, the so-called furanoid linalool oxides (Fig. 10). At an acidic pH ($<3{\cdot}5$) and/or during heat-treatment, the triol **3** had been found to be decomposed to **7** and **8**.[26] In these previous experiments, no formation of the corresponding pyranoid linalool oxides **9** and **10** was observed. We evaluated the hypothesis that linalool oxides are formed from triol **3** using papaya fruit pulp under natural conditions (pH 5·6)

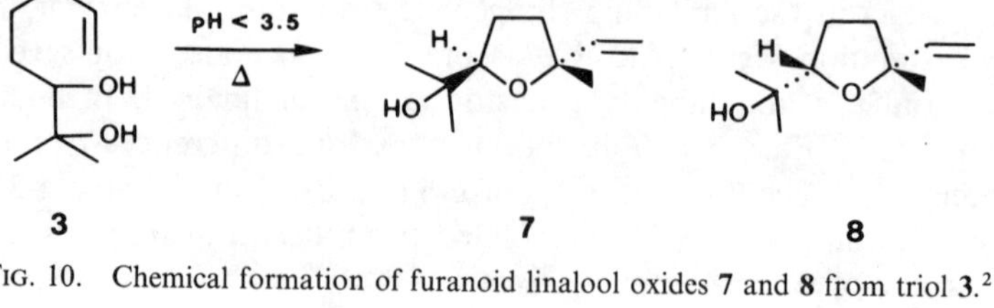

3 7 8

FIG. 10. Chemical formation of furanoid linalool oxides **7** and **8** from triol **3**.[25]

and could find no formation of linalool oxides. Even in model experiments carried out at pH 3·5, only traces of linalool oxides were detected after incubation of **3** at 25°C for three days. As a result of these experiments it is evident that the isomers of 6,7-epoxy-linalool are the natural precursors of linalool oxides as recently proposed by Ohloff *et al.*[40] whose statement was based on earlier findings obtained in a series of chemical reactions as outlined in Fig. 11.

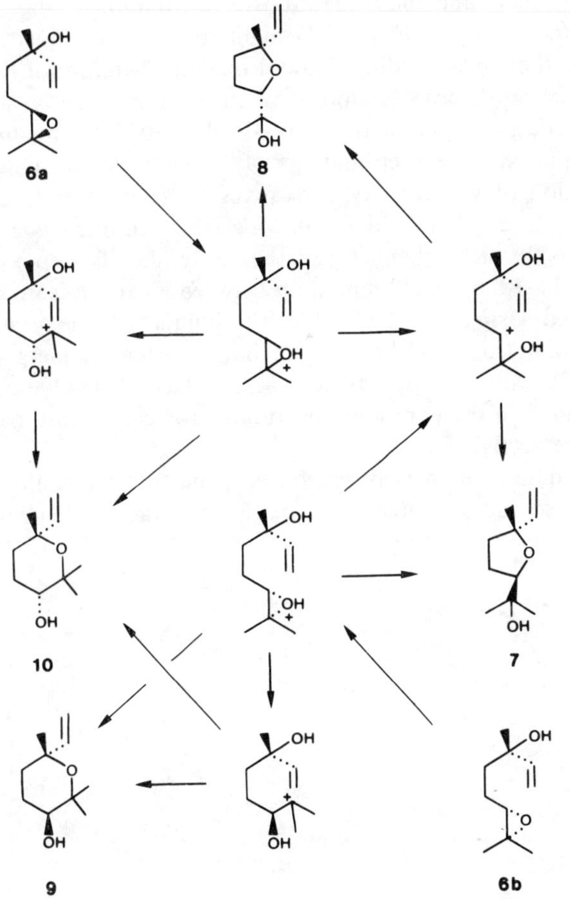

FIG. 11. Chemical formation of isomeric linalool oxides (**7–10**) from 6,7-epoxy-linalool isomers (**6a**/**6b**) obtained from (−)-linalool. Proposal of formation by direct attack of the 3-hydroxyl at the C_6 and C_7 sites of the protonated epoxides or the equivalent carbonium ions.

Epoxy-linalool Oxides

The occurrence of the four isomeric linalool oxides **7–10** in papaya fruit pulp is known from earlier studies.[28,29] In addition to these compounds, the four diastereoisomeric epoxy derivatives of these hydroxy-ethers in their furanoid (**11–14**) and pyranoid forms (**15–18**) were identified in our study. Just as with 6,7-epoxy-linalool (**6**), it was the first time that these epoxy-linalool oxides were detected in a natural plant material. The presence of eight epoxy-linalool oxide isomers and their quantitative distribution in the fruit pulp (e.g. furanoid forms, E:Z = 5:1) might exclude a subsequent epoxidation of the corresponding linalool oxides. Therefore, in Fig. 12, a hypothetical diepoxy-linalool derivative is postulated as a possible biogenetic precursor of the epoxy-linalool oxides. This hypothesis is supported by earlier findings of Osborne,[41] who detected the formation of the diepoxy derivatives of geraniol and nerol after feeding these alcohols to *Pisum sativum* cell cultures.

As to the stereochemistry of the newly identified epoxy-linalool oxides **11–18**, only differentiation between Z- and E-isomers can be provided. Using the HRGC–FTIR technique[42,43] the occurrence of an intramolecular H-bonding in both diastereoisomers indicated Z-configuration. These bands were detected at 3490 cm^{-1} and 3525 cm^{-1} in the pyranoid and furanoid Z-configurational derivatives, respectively.

The data available, as yet, on oxygenated terpene alcohols and derivatives are not sufficient to be generalised; e.g. the formation of

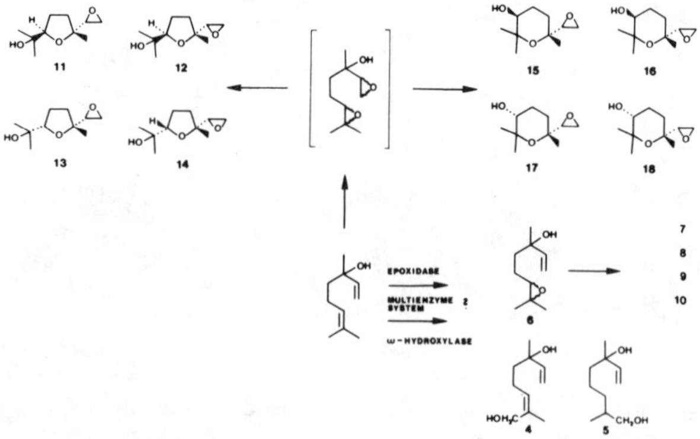

FIG. 12. Proposals for the biogenetic formation of oxygenated linalool derivatives **4–18**.

'polyols' should not yet be considered as a general biochemical pathway of hydrophilisation of terpenoids. However, there is good evidence that our knowledge about the oxygenated terpene alcohols and derivatives is still limited by the techniques commonly used in flavour analysis. Heat treatment and/or acidic conditions generally decompose the labile 'polyols'; epoxy derivatives are unstable too. Therefore special attention has to be devoted to the isolation technique in order to separate the unstable, hydrophilic components from plant tissue.

C_{13}-Norisoprenoids

In fruits and vegetables as well as (in particular) the leaf products tea and tobacco, many norisoprenoid flavour compounds have been found.[40] Several C_{13}-norisoprenoids have also been identified in grapes, wines and distilled alcoholic beverages.[44] The formation of these volatiles has been attributed to the degradation of higher molecular weight terpenoids such as carotenoids.[45] However, the natural precursors of C_{13}-norisoprenoids are, in general, unknown and there are only hypothetical considerations about the pathways by which the norisoprenoid volatiles are formed. Recently, in a series of model studies, Strauss *et al.*[46] demonstrated that, e.g. from megastigma-4-ene-3,6,9-triol and megastigma-4,7-diene-3,6,9-triol, a number of volatile C_{13}-norisoprenoids can be formed due to heat and/or acid treatment. However, these triols have not as yet been observed in plant tissue.

Recently we encountered a similar position, studying the spiro ethers theaspirane A and B from quince (*Cydonia oblonga* Mill.) fruit.[47] As outlined in Fig. 13, the diastereoisomers **19** and **20**, well-known as minor but potent flavour compounds of different plant origins,[40] were identified among the main volatile constituents (Fig. 13(a)) after their careful isolation from the fruit pulp (pH 3·6) by combined high-vacuum distillation/solvent extraction (HVD/SE). Using neutralised fruit pulp (pH 7·0) for flavour isolation, HRGC revealed only traces of **19** and **20** (Fig. 13(b)). After acidification of the fruit pulp up to the natural pH (3·6), isolation by HVD/SE and HRGC analysis again revealed high amounts of **19** and **20**. According to the above-mentioned definitions the theaspiranes A and B have to be considered as 'secondary' flavour compounds, chemically formed from a non-volatile precursor after cell disruption due to the acidic pH of the homogenised tissue.

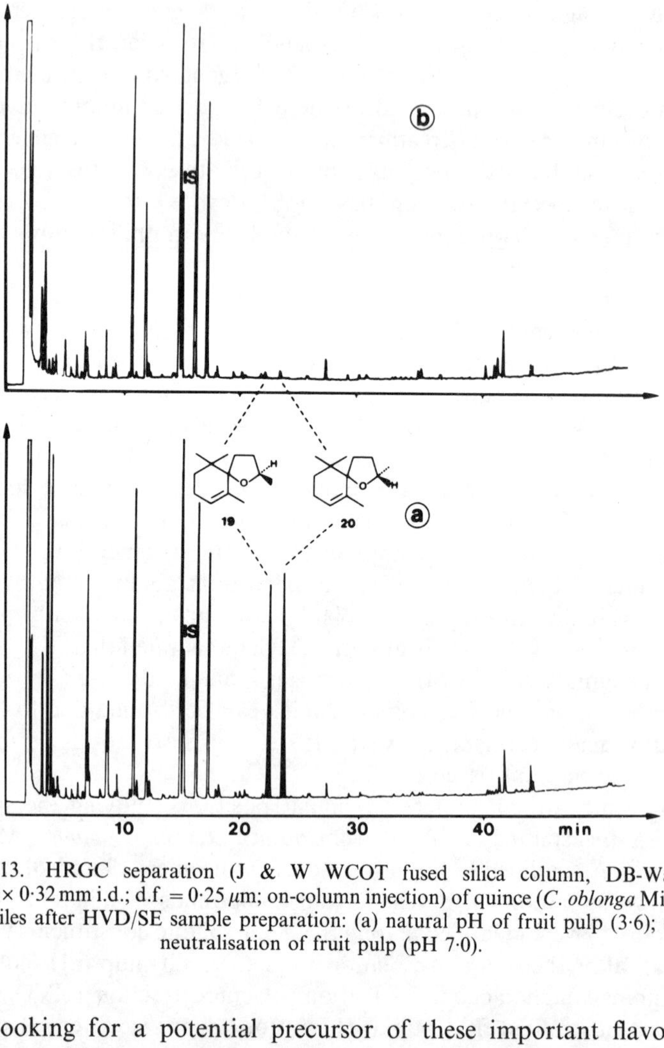

FIG. 13. HRGC separation (J & W WCOT fused silica column, DB-Wax, 30 m × 0·32 mm i.d.; d.f. = 0·25 μm; on-column injection) of quince (*C. oblonga* Mill.) volatiles after HVD/SE sample preparation: (a) natural pH of fruit pulp (3·6); (b) neutralisation of fruit pulp (pH 7·0).

Looking for a potential precursor of these important flavour compounds, we came upon the synthetic work of Schulte-Elte *et al.*[48] carried out on vitispiranes from theaspiranes. Among the different ways to synthesise **19** and **20**, the pathway shown in Fig. 14 seems to be plausible for the biogenetic formation of the theaspiranes. Analogously to the chemical reactions, 1-(2,6,6-trimethylcyclohex-1-enyl)butane-1,3-diol (**21**) can be regarded as a natural precursor of **19** and **20**. As demonstrated by Schulte-Elte *et*

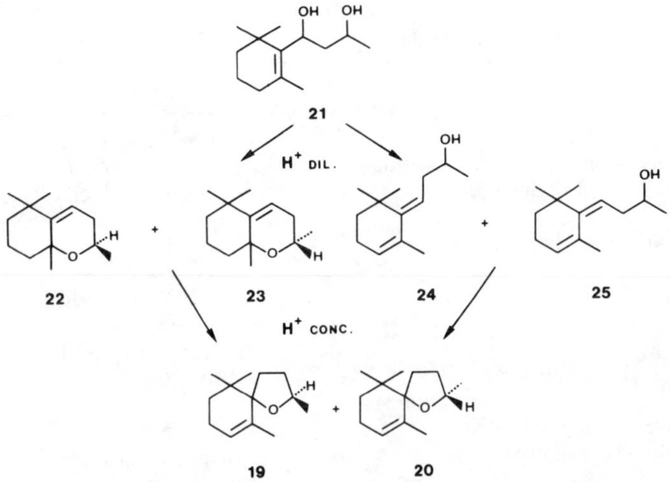

FIG. 14. Chemical formation of theaspiranes **19/20** by strong acid treatment of 1-(2,6,6-trimethylcyclohex-1-enyl)butane-1,3-diol (**21**). Dihydro-edulanes **22** and **23** as well as retro-ionols **24** and **25** are intermediates after weak acid treatment of **21**.

al.,[48] **19** and **20** were formed as stable end-products from **21** on strong acid treatment, whereas the dihydro-edulanes **22** and **23** as well as the retro-ionols **24** and **25** were obtained as unstable by-products on mild acid treatment. However, in our study on quince, the possible precursor **21** could not be detected as yet.

Glycosides

For several years, it has been assumed that the lipophilic terpenoids accumulate e.g. as glycosides in the cytoplasm of plant tissues and, as hydrophilic derivatives, they pass through the membranes.[49–51] However, experimental proof for this hypothesis is still lacking. Except for the iridoids,[23] only in a few cases have genuine C_{10} and C_{15} terpene glycosides been identified. The natural monoterpene glycosides detected in plant tissue until now are outlined in Table 2, as recently summarised by Ishag *et al.*[52] Publications on glycosidic monoterpene derivatives are growing in number, showing the great interest in this class of natural plant constituents.[53] As recently demonstrated by Williams *et al.*,[54] C_{18}-reversed-phase isolates from grapes containing glycosidic derivatives of monoterpenes have been found to be potent precursors of highly

TABLE 2
Natural Monoterpene Glycosides from Plant Tissue[52]

Glycoside	*Species*
Bornyl-β-glucoside	*Ocimum basilicum* L.
Bornyl-monoacetyl-β-D-glucoside	*Psilostrophe villosa*
Citronellyl-β-D-glucoside	*Rosa damascena* cult.-var. *versicolora*
Geranyl-β-D-glucoside	*Rosa damascena* cult.-var. *versicolora*
	Malus silvestris Mill.
	Pelargonium graveolens Ait.
	Tanacetum vulgare L.
Geranyl-β-D-rutinoside	*Vitis vinifera* L.
Geranyl-6-*O*-α-L-arabinofuranosyl-β-D-glucopyranoside	*Vitis vinifera* L.
Isothujyl-β-D-glucoside	*Tanacetum vulgare* L.
Linaloyl-β-rutenoside	*Vitis vinifera* L.
Linaloyl-6-*O*-α-L-arabinofuranosyl-β-D-glucopyranoside	*Vitis vinifera* L.
Menthyl-β-D-glucoside	*Mentha arvensis* var. *piperascens* Hol.
Neoisothujyl-β-D-glucoside	*Tanacetum vulgare* L.
Neomenthyl-β-D-glucoside	*Mentha piperita* L.
Neryl-β-D-glucoside	*Rosa dilecta*
	Pelargonium graveolens
	Tanacetum vulgare L.
Neryl-β-rutinoside	*Vitis vinifera* L.
Neryl-6-*O*-α-L-arabinofuranosyl-β-D-glucopyranoside	*Vitis vinifera*
β-Pinen-10-yl-vicianoside	*Paeonia lactiflora* Pall.

fragrant flavour compounds. Thus, similarly to the 'polyols', they will be of increasing interest for the flavour researcher, in particular biotechnologically.

ACKNOWLEDGEMENTS

Grateful appreciation is expressed to my collaborators D. Katzenberger, V. Lander and P. Winterhalter who performed the studies on oxygenated terpenoids. The financial support of the Deutsche Forschungsgemeinschaft is gratefully acknowledged.

REFERENCES

1. GILDEMEISTER, E. and HOFFMANN, F. (1960). *Die ätherischen Öle*, Akademie-Verlag, Berlin.
2. KUBECZKA, K. H. (Ed.) (1982). *Ätherische Öle*, Thieme, Stuttgart.
3. SCHREIER, P. (1984). *Chromatographic Studies of Biogenesis of Plant Volatiles*, Hüthig, Heidelberg.
4. TRESSL, R., HOLZER, M. and APETZ, M. (1975). In: *Aroma Research*, Eds Maarse, H. and Groenen, P. J., Pudoc, Wageningen, 41–62.
5. FREEMAN, G. G. (1979). In: *Progress in Flavour Research*, Eds Land, D. G. and Nursten, H. E., Applied Science, London, 225–43.
6. SHANKARANARAYANA, M. L., RHAGAVAN, B., ABRAHAM, K. O. and NATARAJAN, C. P. (1982). In: *Food Flavours. A: Introduction*, Eds Morton, I. D. and MacLeod, A. J., Elsevier, Amsterdam, 169–281.
7. BLOCK, E. and SALEEM, A. (1984). *J. Amer. Chem. Soc.*, **106**, 8295–6.
8. BLOCK, E. (1985). *Scientific American*, **252**, 94–9.
9. LARSEN, P. O. (1981). In: *The Biochemistry of Plants*, Vol 7, *Secondary Plant Products*, Ed. Conn, E. E., Academic Press, New York, 501–25.
10. VLIEGENTHART, J. F. G. and VELDINK, G. A. (1982). In: *Free Radicals in Biology*, Vol. 5, Ed. Pryor, W. A., Academic Press, New York, 29–64.
11. GALLIARD, T. (1978). In: *Biochemistry of Wounded Plant Tissues*, Ed. Kahl, G., de Gruyter, Berlin, New York, 155–201.
12. HAMBERG, M. and SAMUELSON, B. (1967). *J. Biol. Chem.*, **242**, 5329–35.
13. EGMOND, M. R., VLIEGENTHART, J. F. G. and BOLDINGH, J. (1972). *Biochem. Biophys. Res. Commun.*, **48**, 1055–60.
14. GROSCH, W. (1982). In: *Food Flavours. A: Introduction*, Eds Morton, I. D. and MacLeod, A. J., Elsevier, Amsterdam, 325–98.
15. KEMP, T. R., KNAVEL, D. E. and STOLTZ, L. P. (1974). *J. Agric. Food Chem.*, **22**, 717–18.
16. KIM, I. S. and GROSCH, W. (1981). *J. Agric. Food Chem.*, **29**, 1220–5.
17. SCHREIER, P. and LORENZ, G. (1982). *Z. Naturforsch.* **37c**, 165–73.
18. HATANAKA, A., KAJIWARA, T., SEKIYA, J. and FUJIWARA, K. (1979). *Agric. Biol. Chem.*, **43**, 175–6.
19. GÖTZ-SCHMIDT, E. M., WENZEL, M. and SCHREIER, P. (1986). *Lebensm. Wiss. Technol.*, in press.
20. STONE, E. J., HALL, R. M. and KAZENIAC, S. J. (1975). *J. Food Sci.*, **40**, 1138–41.
21. DRAWERT, F. (1974). In: *Proc. Colloq. Int. CNRS No. 238*, Ed. Ulrich, M., CNRS, Paris, 309–23.
22. BARDILI, P., MARSCHALL-WEYERSTAHL, H. and WEYERSTAHL, P. (1985). *Liebigs Ann.*, 275–300.
23. ERMAN, W. F. (Ed.) (1985). *Chemistry of the Monoterpenes, Parts A + B, Studies in Organic Chemistry*, Vol. 11, Ed. Gassman, P. G., Dekker, New York.
24. CROTEAU, R. and MARTINKUS, C. (1979). *Plant Physiol.*, **64**, 169–75.
25. WILLIAMS, P. J., STRAUSS, C. R., WILSON, B. and DIMITRIADIS, E. (1985). In: *Topics in Flavour Research*, Eds Berger, R. G., Nitz, S. and Schreier, P., Eichhorn, Marzling, 335–52.

26. Williams, P. J., Strauss, C. R. and Wilson, B. (1980). *Phytochemistry*, **19**, 1137–9.
27. Winterhalter, P., Katzenberger, D. and Schreier, P. (1986). *Phytochemistry*, **25**, 1347–50.
28. Flath, R. A. and Forrey, R. R. (1977). *J. Agric. Food Chem.*, **25**, 103–9.
29. Idstein, H. and Schreier, P. (1985). *Lebensm. Wiss. Technol.*, **18**, 164–9.
30. Takaoka, D. and Hiroi, M. (1976). *Phytochemistry*, **15**, 330.
31. Rapp, A. and Knipser, W. (1979). *Vitis*, **18**, 229–33.
32. Rapp, A., Knipser, W. and Engel, L. (1980). *Vitis*, **19**, 226–9.
33. Engel, K. H. and Tressl, R. (1983). *J. Agric. Food Chem.*, **31**, 998–1002.
34. Devi, J. R., Bhat, S. G. and Bhattacharyya, P. K. (1977). *Indian J. Biochem. Biophys.*, **14**, 359–63.
35. Madyastha, K. M. (1984). *Proc. Indian Acad. Sci.*, **93**, 677–86.
36. Bock, G., Benda, I. and Schreier, P. (1986). *J. Food Sci.*, **51**, 659–63.
37. Behr, D., Wahlberg, T., Nishida, T. and Enzell, C.R. (1978). *Acta Chem. Scand.*, **32b**, 228–34.
38. Suga, T., Hirata, T., Hirano, Y. and Ito, T. (1976). *Chem. Lett.*, 1245–8.
39. Tschesche, R., Ciper, F. and Breitmaier, E. (1977). *Chem. Ber.*, **110**, 3111–17.
40. Ohloff, G., Flament, I. and Pickenhagen, W. (1985). *Food Rev. Int.*, **1**, 99–148.
41. Osborne, M. J. (1979). In: *Terpenoid Biosynthesis in Tissue Cultures of Jasmine*, Ph.D. thesis, University of London.
42. Schreier, P. and Idstein, H. (1985). *Z. Lebensm. Unters.-Forsch.*, **181**, 183–8.
43. Fröhlich, O., Kahre, C. and Schreier, P. (1986). *Z. Anal. Chem.*, **324**, 314.
44. Nykänen, L. and Suomalainen, H. (1983). In: *Aroma of Beer, Wine and Distilled Alcoholic Beverages*, Akademie-Verlag, Berlin.
45. Enzell, C. (1985). *Pure Appl. Chem.*, **57**, 693–700.
46. Strauss, C. R., Dimitriadis, E., Wilson, B. and Williams, P. J. (1986). *J. Agric. Food Chem.*, **34**, 145–9.
47. Winterhalter, P., Lander, V. and Schreier, P. (1986). In prep.
48. Schulte-Elte, K. H., Gautschi, F., Renold, W., Hauser, A., Fankhauser, P., Limacher, J. and Ohloff, G. (1978). *Helv. Chim. Acta*, **61**, 1125–33.
49. Amelunxen, F. (1965). *Planta Med.*, **13**, 457–73.
50. Heinrich, G. (1973). *Planta Med.*, **23**, 154–65.
51. Hörster, H. and Skopp, K. (1976). *Planta Med.*, **29**, 208–15.
52. Ishag, K. E. A., Jork, H. and Zeppezauer, M. (1985). *Z. Anal. Chem.*, **321**, 331–6.
53. Paulsen, H., Le Nguygen, B., Sinnwell, V., Heemann, V. and Seehofer, F. (1985). *Liebigs Ann.*, 1513–36.
54. Williams, P. J., Strauss, C. R., Wilson, B. and Massy-Westropp, A. (1982). *J. Agric. Food Chem.*, **30**, 1219–23.

7

Plant Cell Cultures and the Production of Food Flavour

M. J. C. Rhodes

AFRC Institute of Food Research, Norwich, UK

ABSTRACT

Significant progress has been made towards the development of systems to make food flavour compounds using the approach of cell immobilisation coupled with the use of continuous flow fermenter systems. There is a need to make further improvements in the productivity and stability of the cell lines to be used in such processes and in the rate at which they release product into the medium, which is the effective production rate of the fermenter system. We have operated such systems for extended periods but more needs to be done to define the nutritional requirements for maintaining cells in the productive phase for extended periods. There are many problems to be solved but there is every prospect that biotechnological approaches to food flavour production may ultimately find commercial application.

INTRODUCTION

There has been substantial progress towards the development of biotechnological processes to make industrially important products using plant cells in culture. The first commercially successful process is used to make the anti-inflammatory drug shikonin by fermentation of cells of *Lithospermum erythrorhizon*. Most work in the area is directed towards drugs such as the *Catharanthus* alkaloids, vincristine and vinblastine, and the Papaver alkaloids, codeine and morphine, or drug precursors such as diosogenin from *Dioscorea*, which are both expensive (more than 500–1000 US$/kilo) and in

TABLE 1
Cost of some Food Flavour Compounds which are Targets for Biotechnological Processes

Product	*Plant species*	*Value (US$/kilo)*
Capsaicin	*Capsicum frutescens*	30–50
Crocin	*Crocus sativa*	1000
Humulone	*Humulus lupulus*	50
Mint oils	*Mentha* sp.	30
Quassin	*Quassia amara*	5–7000
Vanillin	*Vanilla planifolia*	30

short supply.[1] In contrast, the amount of effort which has been devoted to the production of food flavour compounds is small. This is largely because food additives, with a few exceptions, are relatively inexpensive and are thus less attractive targets (see Table 1). The exceptions include the bittering agent quassin (from *Quassia amara*) at about 5000 US$/kilo and the pigment saffron (from *Crocus sativa*) at about 1000 US$/kilo. Typically the food additives which have sufficiently large markets to make their biotechnological production viable have values in the range of 30–80 US$/kilo. If biotechnological processes to produce flavours are to have a major impact in the food industry, it is necessary to extend the technology to these lower priced products.

Some of the work which has been done on food flavour production in cell culture is summarised in Table 2 and includes studies on a range of different flavour compounds from alkaloids such as quinine, and phenolics such as capsaicin, to terpenes such as the mint oil monoterpenes. In the case of the mint oils, many cultures produce a different spectrum of monoterpenes from that of the intact tissue.[2] In onion the culture fails to make the characteristic flavour compounds even though levels of the enzyme active in their formation, allinase, are similar in culture to those of the intact tissue. The lack of flavour generation is due to the complete absence of the major flavour precursor, *S*-prop-1-enyl-L-cysteine sulphoxide, even though traces of the related compound *S*-methyl-L-cysteine sulphoxide are formed.[3] Thus, the culture contains some but not all of the enzymes required for precursor generation. In celery, the formation of the phthalide derivatives, which are important factors in celery flavour, is dependent on aggregation and differentiation

TABLE 2
Some studies on Food Flavour Compounds in Cell Culture

Plant	*Target product*	*Type of culture*[a]	*Comments*	*Ref.*
Allium cepa (onion)	Onion aroma	C	Enzyme allinase present at normal levels. Major precursor *S*-t-prop-l-enyl-L-cysteine sulphoxide absent. Trace of *S*-methyl-L-cysteine sulphoxide.	3
Apium graveolens (celery)	Phthalide flavour cpds	C/CSC	3-Isobutylidene-3α,4,5,6-tetrahydrophthalide and 3-isobutylidene-3α,4-dihydrophthalide formed in differentiated cultures.	4
			Increase in phthalide production, degree of aggregation and greening on transfer from 2,4-D to 3,5-D containing media.	5
Capsicum frutescens (chilli pepper)	Capsaicin	CSC/ICC	*De novo* synthesis and biotransformation from iso capric acid. Stimulation of production on immobilisation.	6
Cinchona ledgeriana	Quinine	CSC/ICC	Cultures produce cinchonine and cinchonidine in addition to quinine and quinidine.	15
Mentha sp. (mint)	Mint oil components	C/CSC/ICC	Triterpene acids, oleanolic, ursololic, pomolic, maslinic, Callus rich in pulegone and menthofuran/cf. plant rich in menthone/menthol.	2

[a] C = callus; CSC = cell suspension culture; ICC = immobilised cell culture.

within the culture.[4] Watts *et al.*[5] have shown that transfer from one type of auxin to another [from 2,4-dichlorophenoxyacetic acid (2,4-D) to the related isomer (3,5-D)] acts as a switch to induce phthalide formation. Lindsey and Yeoman[6] have demonstrated the *de novo* formation of the pungent principle of chilli pepper, capsaicin, in cultures of *Capsicum frutescens*, and shown that immobilisation of cells within a polyurethane matrix can have a marked stimulatory effect on the formation of this flavour compound.

Work at the IFR, Norwich, has concentrated on flavours which depend on single, well-characterised molecules rather than complex mixtures, such as the essential oils, because of the greater ease with which such systems may be manipulated in culture. As a way of extending the technology to the relatively inexpensive food flavours, we have worked towards the development of fermenter systems based on immobilised plant cells operating in a continuous mode. If products are released from the cells, the process can operate with continuous collection of the product without having to harvest the cells. Thus, extended use may be obtained from a given biomass volume with consequent improvements in the economics of the process. The approach involves both the development of cell lines with appropriate properties, i.e. high stable production of the compounds of interest, together with high and sustainable rates of release of the product from the cells to the medium, and the development of suitable immobilisation and fermentation technologies with which to exploit such cells.

TYPES OF PLANT CELL CULTURES

Before discussing these two aspects it is necessary to describe the main types of plant cell culture now in use. Cultures of plant cells are developed by incubating sterile sections of plant material (leaves, stems, etc.) on a semi-solid medium containing a full range of nutrients and growth promoting plant hormones, particularly auxins and cytokinins. An outgrowth of disorganised cells, the callus, develops at the cut surfaces of the plant material and can eventually be excised and grown under sterile conditions independent of the parent plant tissue. The callus can be dispersed in liquid media and established as a cell suspension culture. Most work on secondary products such as drugs and flavours in cell cultures has been done with suspension cultures, but there is increasing interest in immobilised cell cultures for the reasons noted above. Cells from suspension culture may be immobilised in beads of calcium alginate gel,[7] or used to inoculate semi-rigid materials such as polyurethane foam or nylon.[8] More recently there has been interest in organised culture such as the fast growing 'hairy' root cultures[9,10] developed by transformation of plant cells by the bacterium *Agrobacterium rhizogenes*. These roots produce a normal spectrum of secondary products and their study for the production of root-derived drugs is under investigation. However, there are no reports so far of their application to food flavour production.

MANIPULATION OF SECONDARY METABOLISM IN CULTURE TO IMPROVE PRODUCT YIELD

Most of the methods used to manipulate the yield of commercially important products from plant cells in culture are essentially empirical since our knowledge of the enzymology and molecular genetics of the development, operation and regulation of the metabolic pathways leading to such products is often very limited. The procedure devised by Zenk *et al.*[11] is widely used. It involves selecting a range of high producing plants to initiate a number of cell lines and the development of separate growth and production media. The approach leads to the development of two-stage processes. The first is a growth stage in which cells are grown in a medium optimised to give maximal growth; the second is a production stage in which cells are all transferred to a second medium giving maximal product formation. Often the formation of secondary products such as flavour compounds occurs in cells which have ceased cell division and expansion; thus the two-stage operation is compatible with this natural separation of the growth and production. Plant cell culture media usually have sucrose to support heterotrophic growth and use either NO_3 or NH_4^+ as the N source supplemented by macronutrients and micronutrients together with vitamins such as inositol and nicotinic acid which plant cells cannot produce in sufficient quantities in culture. Growth and productivity can be regulated by manipulation of factors such as the type and concentration of the carbon and nitrogen sources, of the ratio of carbon/nitrogen in the medium and by the level of inorganic phosphate.[12] In addition, plant growth substances are essential components of plant cell culture media. The type and absolute and relative concentrations of these compounds affect the rate of growth and the morphology of the culture (degree of aggregation and extent of differentiation into organised structures) as well as the production of secondary products. They are very powerful tools in the manipulation of cells in culture but their effects on secondary metabolism are not necessarily direct and this, to some extent, limits their usefulness. Change from one hormonal regime to another may induce expression of the enzymes involved in the pathway but this may be associated with undesirable morphological changes.

A further element of Zenk's approach relates to the genetic heterogeneity of plant cells in culture. The special conditions in culture, i.e. the presence of potentially mutagenic levels of synthetic auxins and the breakdown of organisation between cells, promotes a high degree of genetic instability with poly- and aneuploid cells commonly present. The screening of such heterogeneous cultures can lead to the isolation of variant cell types with enhanced productive capacity. The screening process involves cloning individual cells (or protoplasts) to produce small colonies of cells which can be analysed for the presence of the product at an early stage. This procedure is most easily applied to coloured products which may be screened visually but can be applied to other compounds if highly sensitive analytical procedures such as immunoassays (RIA or ELISA) are available. Since reversion of productive capacity can occur, it is often necessary to repeat the screening process through a number of cycles to isolate cell lines which have both high and stable production rates.

These empirical approaches have led to major improvements in the productivity of some cell lines. In *Lithospermum* the production of shikonins has been raised from the level of about 0·8% of the dry weight in the plant roots, from which it is normally extracted, to over 23% of the dry weight in cultures which have been developed using the Zenk procedure.[13] There are many examples of cells in culture which accumulate over ten-fold more of a product than the level in the parent plant tissue.[14] In spite of this success there are a number of desirable products which have yet to be found in cell cultures. Prime examples are the dimeric *Catharanthus* alkaloids, vinblastine and vincristine. One of the precursors of the dimer, catharanthine, can accumulate in levels in culture in excess of that found in the plant. However, the cultures lack the enzyme catalysing the final steps in the biosynthesis of the other precursor, vindoline, and the enzyme which catalyses the condensation of the two monomers. This fragmentary expression of secondary pathways is common in culture and we need to devise ways of more effectively controlling expression of the relevant genes in culture.[12]

More rational approaches to these problems require much more knowledge of the biochemistry and genetics of the biosynthetic sequences in question. The possibility of transferring whole pathways from higher plants to micro-organisms may ultimately be desirable but is beyond present gene technology. It may, however, be possible to move copies of genes coding for single enzymes under the influence of high expression promoters into a plant cell line deficient in that activity and in this way repair its biosynthetic deficiency. Thus work to isolate and characterise the genes coding for key enzymes, both those which have a regulatory role in the metabolic sequence or those which are very poorly expressed in culture, is important in developing more direct routes for regulating expression of secondary pathways in cultures.

EXPLOITATION OF PRODUCTIVE CELL LINES

The approach used at IFR, Norwich, to exploit plant cells is illustrated in Fig. 1. It involves the immobilisation of an inoculum of cells into particles (8 × 8 × 8 mm) of wash reticulated polyurethane foam (Rhodes *et al.*, unpublished data). The process is designed to minimise the use of free cells, to inoculate a defined volume of the intraparticle space sufficient to initiate active growth and to give

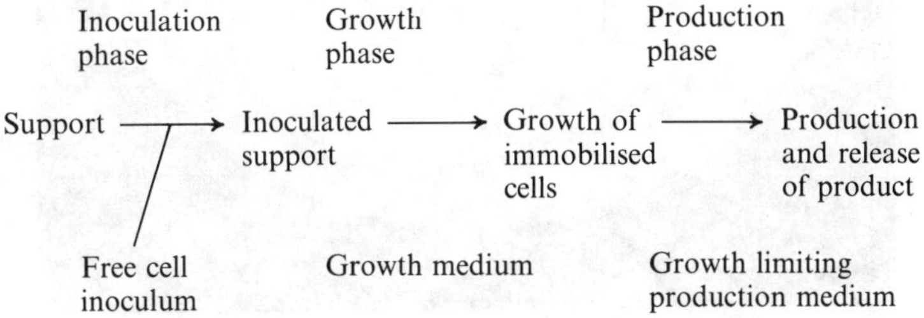

FIG. 1. The approach used at IFR, Norwich, for exploitation of immobilised plant cells for food flavour production.

evenness of loading between individual foam particles. The reticulated foam (Fig. 2) represents a very open structure (~97% void) and the aim is to achieve an initial 5–10% occupancy of the space, to grow the cells *in situ* to a high density (up to 80–100% of the available volume occupied) and finally to exploit these immobilised cell particles in a growth-restricting productive phase (Fig. 3). An essential feature of this approach is that the product is released from the cells into the medium from which it can be recovered without the need to harvest the biomass. This approach, using immobilised cells in extended continuous operations, can have a favourable effect on the economics of production compared with conventional batch fermenter operations.

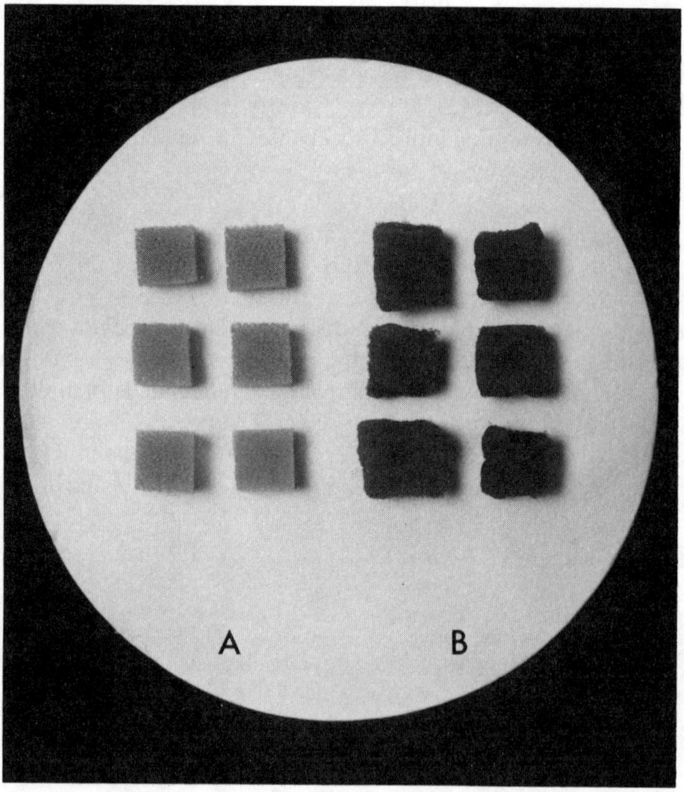

FIG. 2. Polyurethane foam particles: (A) before inoculation; (B) fully loaded with cells of *Cinchona pubescens*.

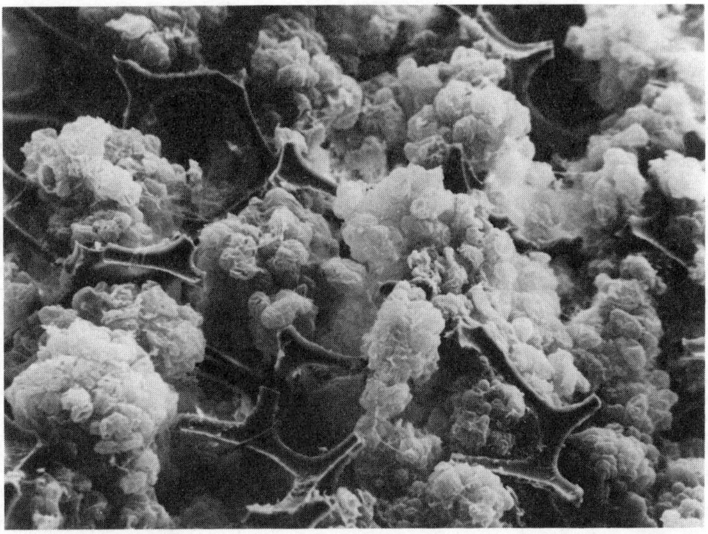

FIG. 3. Scanning electron micrograph of cells of *Beta vulgaris* immobilised in polyurethane foam. Original magnification ×40.

Air-stirred fermenters have been designed at IFR, Norwich, to exploit plant cells immobilised on polyurethane particles. Growth and production media are developed by determining optimal combinations of hormones and levels of nutrients. In the productive phase, growth restriction can be achieved by lowering the level of essential nutrients (carbon or nitrogen source, phosphate levels etc.) to limiting values, decreasing hormone levels or by the use of osmotic stress (hypertonic levels of mannitol). The maintenance of intact viable cells for extended periods is relatively easy to achieve but in some instances it is more difficult to determine the nutritional requirements for extension of the productive phase of the culture.

For many of the products which are currently targets for biotechnological investigation, there is a mechanism in the intact plant by which they are moved from the site of synthesis to the site of deposition which involves transmembrane transport. It is likely that this transport mechanism may be expressed in culture under appropriate conditions. In the case of some *Cinchona* cultures,[15] up to 80% of the alkaloids produced by the cells are released into the medium. The mechanism has not yet been completely elucidated,[16] but at least in part it involves an ion-trapping mechanism in which uncharged alkaloid molecules cross the cell membranes and are trapped within acidic components such as vacuoles in a charged form. However, this may not be the only mechanism operating, and there is interest in the recent findings of Deus-Neumann and Zenk[17] that an active ATP-dependent process involving highly selective carriers is responsible for the movement of some alkaloids across the tonoplast membrane. Whether such a mechanism also operates with quinine in *Cinchona* remains to be tested. It is possible to effect release of products, not transported out of the cells and normally stored in the vacuole, by treatments with chemicals which permeabilise the cell membranes. However, with vacuolar products it is necessary to permeabilise both the tonoplast and the plasmalemma to release the product. This process leads to the flooding of the cytoplasm with toxic vacuolar constituents and frequently leads to damage and ultimately to cell death.[18]

The stability of products in the culture medium after release from the cells can be a problem. Of particular importance is the enzymic degradation of the product. One example is the rapid oxidation of humulone, the precursor of the bitter principle of hops, in cultures of *Humulus* which has been shown to be due to an extracellular peroxidase in the presence of H_2O_2 generated and released by the cells. Very high rates of humulone oxidation (in excess of 200 pmol/s/g.fr.wt) have been observed,[19] and it is necessary either to select cell lines with low levels of peroxidase or to devise ways of removing and stabilising the released product to prevent its oxidation. Peroxidase is known to be released from many different cell lines in culture[20] and, as it has such a broad specificity for acceptor molecules, oxidation of extracellular plant products by peroxidase is likely to be a problem in many different systems.

As an approach to this problem we have used adsorbents to rapidly remove extracellular products. The Amberlite XAD macroreticular resins have been shown to be useful with products such as anthraquinones and the tobacco alkaloids. XAD-7 has been used to remove anthraquinones from culture of *Cinchona ledgeriana*. These are produced by some cell lines in addition to the quinine alkaloids with which they share a common precursor, chorismic acid. The anthraquinones are all nearly completely released from the cells into the medium where, because of their low water-solubility, they precipitate.[21] They are also phytotoxic and it was thus desirable to remove them by adsorption to relieve this toxicity.

The interesting finding was that XAD-7 at appropriate concentrations not only adsorbed all the extracellular anthraquinone but gave an overall stimulation of anthraquinone formation which can be 15–20 times greater than that in culture in the absence of adsorbent.[22] This effect of resin may be due to either (i) adsorption and protection of the product from degradation or (ii) adsorption of factors responsible for feedback regulation of the biosynthetic pathway and thus removing the regulatory constraints on the pathway. In this particular case it is very likely that the second mechanism is operational but in other cases the role of the adsorbent may be to protect against degradation. The rapid removal of the product by adsorption may also stimulate transport by reducing the effective external concentration to zero. These studies demonstrate the value of adsorbents in protecting the product from degradation and in promoting transport and biosynthetic processes. The XAD resins are not ideal since they can adsorb medium components such as vitamins and growth substances which may be detrimental to the growth.[22] It is necessary to develop more selective non-toxic adsorbents to maximise the benefits of rapid removal of the product.

REFERENCES

1. Balandrin, M. F., Klocke, J. A., Wirtele, E. S. and Bollinger, W. H. (1985). *Science*, **228**, 1154–60.
2. Bricout, J. and Paupardin, C. (1974). *Compt. Rend. Acad. Sci.*, **278**, 719–22.
3. Selby, C., Galpin, I. J. and Collin, H. A. (1979). *New Phytol.*, **83**, 351–9.
4. Al-Abta, S., Galpin, I. J. and Collin, H. A. (1979). *Plant Sci. Lett.*, **16**, 129–34.
5. Watts, M. H., Galpin, I. J. and Collin, H. A. (1985). *New Phytol.*, **100**, 45–56.
6. Lindsey, K. and Yeoman, M. M. (1984). *Planta*, **162**, 495–501.
7. Brodelius, P. (1983). In: *Immobilised Cells and Organelles*, Ed. Mattiasson, B., CRC Press, Boca Raton, 27–55.
8. Rhodes, M. J. C. (1985). In: *Topics in Enzyme and Fermentation Biotechnology*, Vol. 10, Ed. Wiseman, A., Ellis Horwood, Chichester, 51–87.
9. Flores, H. E. and Filner, P. (1985). In: *Primary and Secondary Metabolism of Plant Cell Cultures*, Eds Neumann, K.-H., Barz, W. and Reinhard, E., Springer-Verlag, Berlin, 174–85.
10. Hamill, J. D., Parr, A. J., Robins, R. J. and Rhodes, M. J. C. (1986). Secondary product formation in cultures of *Beta vulgaris* and *Nicotiana rustica* transformed with *Agrobacterium rhizogenes*, *Plant Cell Reports*, **5**, 111–14.

11. ZENK, M. H., EL-SHAGI, H., ARENS, H., STOCKIGT, J., WEILLER, E. W. and DEUS, B. (1977). Formation of indole alkaloids, serpentine and ajmalicine in cell suspension cultures of *Catharanthus roseus*. In: *Plant Tissue Culture and its Biotechnological Application*, Eds Barz, W., Reinhard, E. and Zenk, M. W., 27–43.
12. RHODES, M. J. C., ROBINS, R. J., HAMILL, J. and PARR, A. J. (1986). *New Zealand J. Technol.*, in press.
13. FUJITA, Y., TAKAHASI, S. and YAMADA, Y. (1984). *Proc. 3rd Europ. Conf. Biotechnol.*, **1**, 161–6.
14. FOWLER, M. W. (1981). *Chem. Ind.*, 4 April, 229–53.
15. ROBINS, R. J., PAYNE, J. and RHODES, M. J. C. (1986). *Planta Med.*, 220–6.
16. PARR, A. J., ROBINS, R. J. and RHODES, M. J. C. (1986). *Physiol. Veg.*, in press.
17. DEUS-NEUMANN, B. and ZENK, M. W. (1986). *Planta*, **167**, 44–53.
18. PARR, A. J., ROBINS, R. J. and RHODES, M. J. C. (1984). *Plant Cell Reports*, **3**, 262–5.
19. ROBINS, R. J., FURZE, J. M. and RHODES, M. J. C. (1985). *Phytochemistry*, **24**, 709–14.
20. STICHER, L., PENEL, C. and GREPPIN, H. (1981). *J. Cell Sci.*, **48**, 345–53.
21. ROBINS, R. G., PAYNE, J. and RHODES, M. J. C. (1986). *Phytochemistry*, in press.
22. ROBINS, R. J. and RHODES, M. J. C. (1986). *Appl. Microbiol. Biotechnol.*, **24**, 35–41.

8

Liquid Carbon Dioxide Extraction of Flavour Materials

D. A. Moyler

Pauls Flavours & Fragrances Ltd, Reigate, Surrey, UK*

ABSTRACT

The Pauls company were the first commercial extractors to offer a range of essential oils prepared using liquid carbon dioxide. This paper summarises those developments and demonstrates their specific advantages, from a technical viewpoint, over conventionally prepared essential oils. A comparison of the product characteristics and yield obtained, from the same batch of botanical, by liquid carbon dioxide and steam distillation, is made. In particular, gas–liquid chromatography using the standardised 'retention index' system of calibration will be used to identify and compare the quantities of components. Areas of application to developments and improvements in food flavours are suggested, especially for the range of oils currently available. The paper reviews the parameters of all carbon dioxide extractions, whether sub- or super-critical, and suggests possibilities of further developments in essential oil technology applied to the flavour and fragrance industry.

INTRODUCTION

When selecting an ingredient for flavouring food, the best comes naturally. The current consumer preference, which borders on demand, for additive free products and the dislike of E numbers on product listings, causes many dilemmas for flavourists.

Carbon dioxide is natural and pure. At high pressure and temperatures below 31°C it liquefies to become a non-polar, selective

* Now trading as Felton Worldwide.

solvent which is odourless, tasteless, colourless, easily removed and non-flammable—the ideal safe solvent. It can be used in two modes: (i) hot and supercritically as a fluid, to dissolve caffeine from coffee, tea and kola and to extract oleoresins; (ii) cool and subcritically as an extractant for essential oils under conditions which do not cause degradation.

NATURAL INGREDIENT CHOICE AND MERITS

Raw Botanical

Any creative flavourist formulating a new food product should never forget the basic option of raw or chopped botanical in food. It is usually weak compared to extracts but it is visually interesting and cheap.

Steam Distillation

The ubiquitous techniques used for the preparation of the volatile aromatic components of botanicals have several disadvantages. These have been documented independently of this author's work.[1] The oils produced suffer from loss of the highly volatile and some water-soluble components. In addition unwanted offnotes are generated during distillation. These will age and mature if the oil is left to mature but this maturation can take varying lengths of time. This is inconvenient and incurs the cost of storage as well as delaying the oil's availability for use. These 'still notes' are not the only degradation products of steam distillation. Some of the monoterpene hydrocarbons found in steam-distilled oils were formed during the processing as explained later in this paper.

Molecular weights of about 200 will distil with steam at atmospheric pressure, and this is only slightly increased by the use of steam pressure as in the hydrodiffusion method of oil extraction.

Solvent Extracts

This is a popular method for making oleoresins, absolutes, tinctures and concretes (except orris, which is steam distilled). These products contain solvent residues; the lower the quantity of residual solvent, the more is the loss of the volatile oil topnote. The removal of solvent is usually aided by heat or steam which does cause some degradation of the extract. Any moisture present in the botanical will steam out of the extract with the solvent causing co-distillation of topnotes. Unselective solvents also extract lipids, waxes, tannins and chlorophyll when present, which are usually unwanted, particularly in water-based foods.

Commonly used solvents are hexane, acetone, ethanol and ethyl acetate.

According to the phase diagram (Fig. 1), carbon dioxide can be used as an extractant in two phases, supercritically and subcritically, the critical point being 31°C and 80 bar. At temperatures and pressures in excess of the critical point, CO_2 is in the supercritical fluid phase, and below that point it can exist as a liquid.

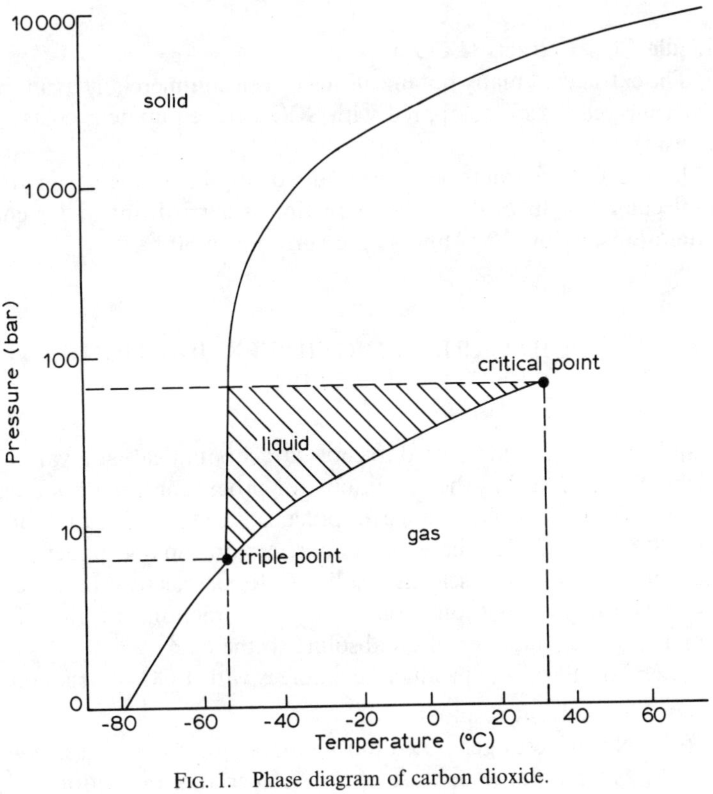

FIG. 1. Phase diagram of carbon dioxide.

Let us consider the practical implications of the profile of the extracts and equipment required to produce them.

Supercritical CO_2 Extracts (SCO_2)

At present SCO_2 extracts of spices are not commercially available and are unlikely to become available in the near future. The reasons are twofold, lack of demand and high cost of production. SCO_2 extracts are non-selectively dissolved from the botanical and are similar to the oleoresin produced by conventional solvents, but solvent-free without residues. Conditions of practical use are 60–80°C and 300–700 bar. At these operating parameters SCO_2 has been used commercially to decaffeinate coffee and tea for a number of years. It is, in the author's opinion, unsurpassed in the quality of the decaffeinated beverage compared to any other solvent.

Liquid CO_2 Extracts (LCO_2)

The extracts of many botanicals have been commercially available for four years and, compared with SCO_2, the equipment costs are considerably less.

Liquid CO_2 extracts selectively, dissolving components up to a molecular weight of 400. The extraction is carried out under cool conditions, below 10°C and at pressures up to 80 bar.

LIQUID CARBON DIOXIDE EXTRACTION DEVELOPMENT

Liquid carbon dioxide is used to cool extract botanicals as a variable technique, optimising the conditions for processing each botanical by experiment. The usual range for practical extraction is 0 to +10°C and 8–80 bar. Under these conditions carbon dioxide liquefies to become a non-polar selective solvent. It is selective because it extracts the essential oil plus the lighter fractions of the resin imparting the character of an absolute to the oil.[2]

There are five clear product advantages with LCO_2 extraction:

(i) No solvent residues
(ii) No offnotes generated during extraction
(iii) More topnote because of low-temperature extraction

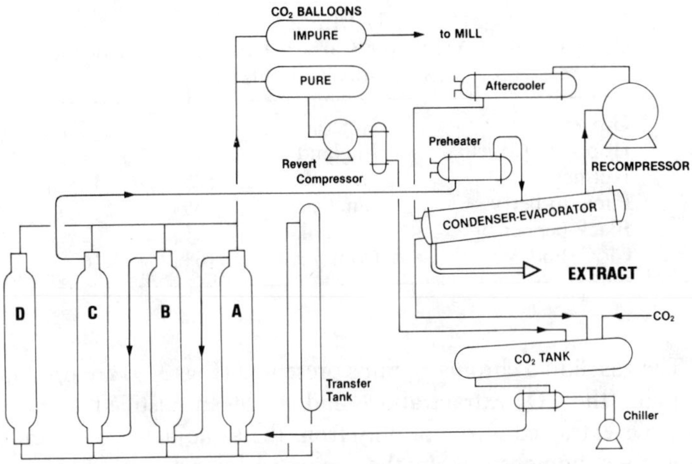

FIG. 2. Extraction circuit.

(iv) More backnote because CO_2 extracts the character of the absolute
(v) Better solubility because no terpenes are generated during extraction

The extracts resemble the raw botanical more closely than do the steam-distilled oils.

Figure 2 shows the extraction circuit used.[3] Some physical constants for CO_2-extracted and steam-distilled products are compared in Table 1, and yields are compared in Table 2.

TABLE 1
Comparison of Physical Constants for Carbon Dioxide-extracted and Steam-distilled Oils

Oil	*S.G. at 20°C*		*R.I. at 20°C*	
	CO_2	*S.D.*	*CO_2*	*S.D.*
Ginger	0·960	0·880	1·510	1·492
Hop	0·880	0·880	1·478	1·488
Nutmeg	0·890	0·864	1·470	1·473
Pimento berry	1·010	1·030	1·520	1·530
Black pepper	0·885	0·878	1·485	1·485
Clove bud	1·040	1·050	1·530	1·532

TABLE 2
Comparison of Yields for Carbon Dioxide-extracted and Steam-distilled Oils

Ginger	Nigeria	3%	3%
Hop var. Target	England	1%	1%
Nutmeg	Grenada	12%	13%
Pimento berry	Jamaica	5%	5%
Black pepper	Sarawak	2%	2%
Clove bud	Comores	16%	16%

The gas–liquid chromatograms presented (Figs 3–9) are produced in pairs, the CO_2 extract above and the steam-distilled oil below. The oils extracted were not only from the same batch but even the same sack, homogenised by the coning and quartering technique of sampling. The percentage of each component is presented on the 'y' axis and the standardised temperature-programmed Retention Index run on polarity tested columns using the NC criteria.[4–7] All these oils, and many more besides, show at least some of the five characteristic features outlined earlier.

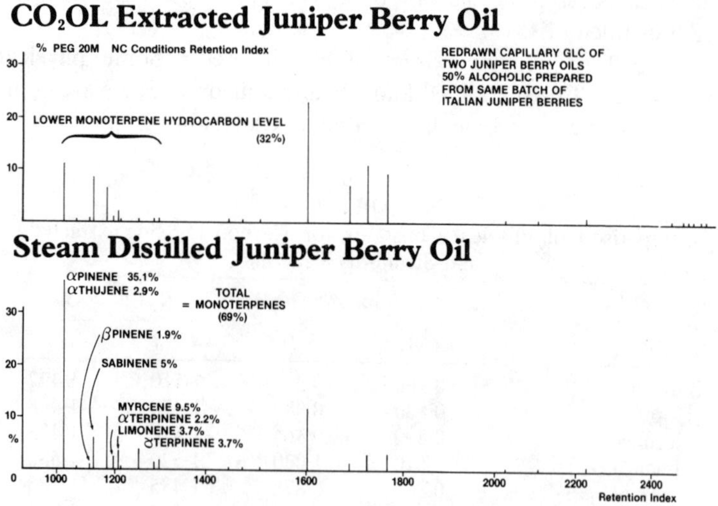

FIG. 3. Comparison by GLC of juniper berry oils.[2]

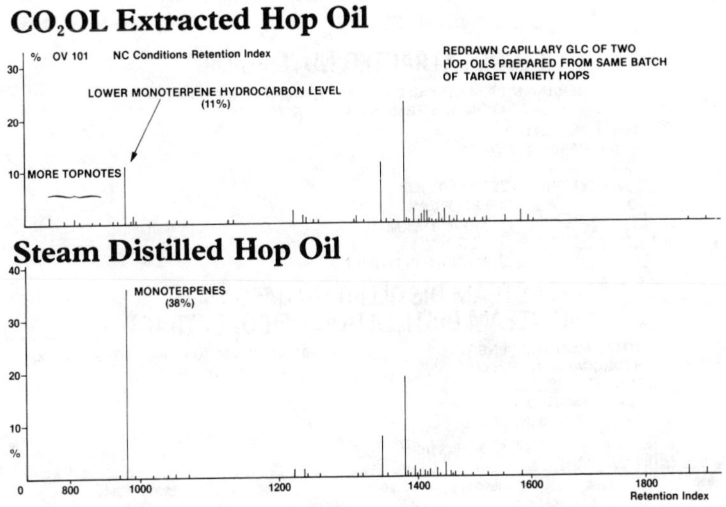

FIG. 4. Comparison by GLC of hop oils.[2]

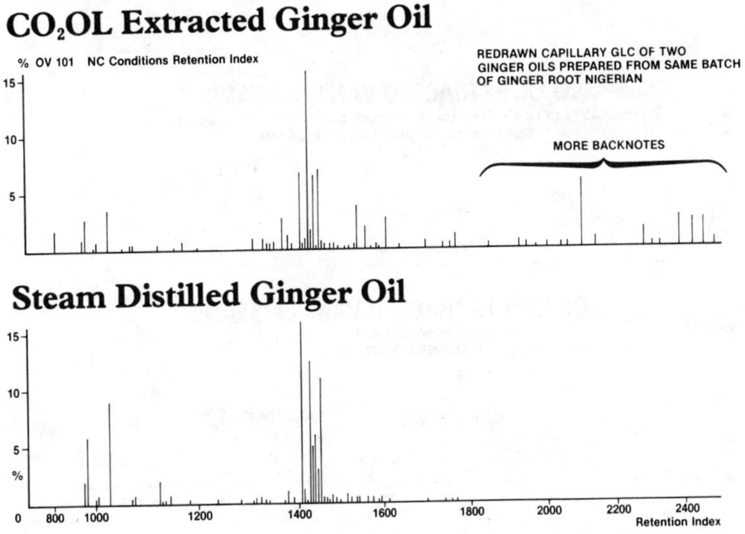

FIG. 5. Comparison by GLC of ginger oils.[2]

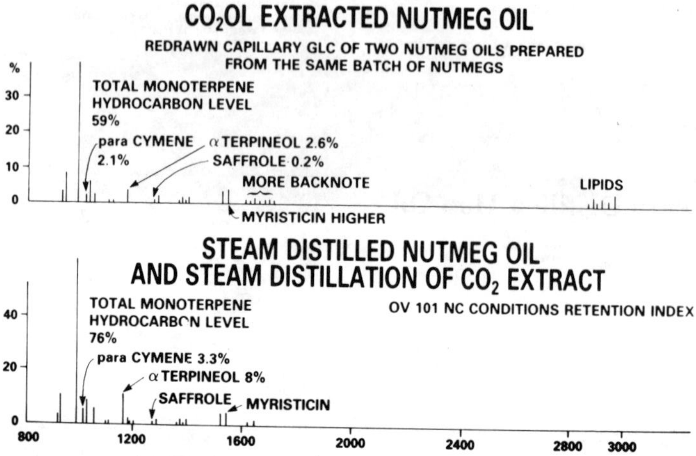

FIG. 6. Comparison by GLC of nutmeg oils.[8]

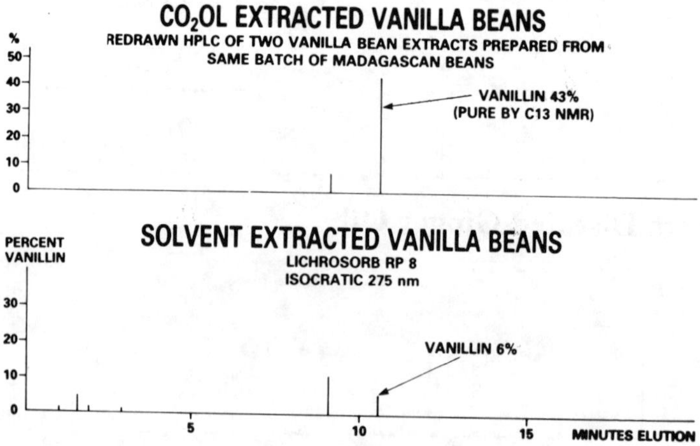

FIG. 7. Comparison by HPLC of vanilla bean extract.[8]

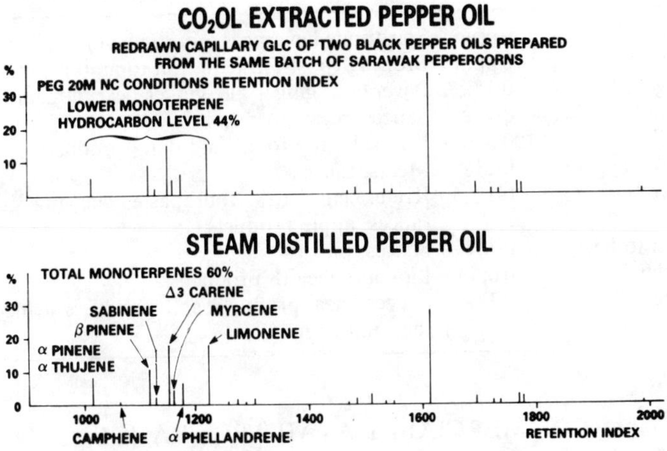

FIG. 8. Comparison by GLC of black pepper oils.[8]

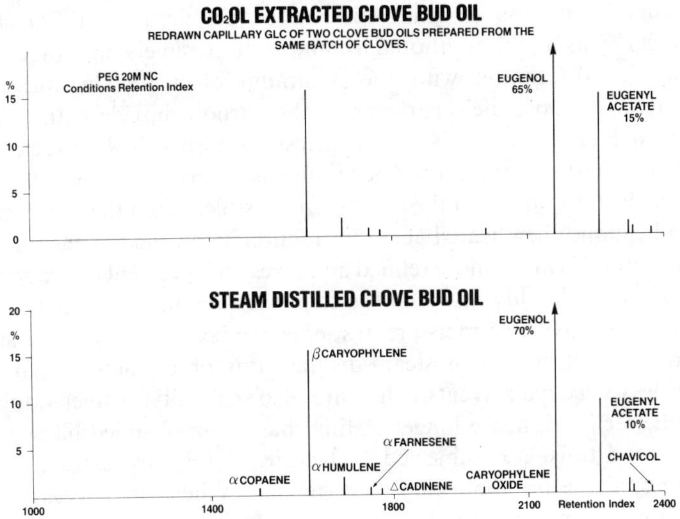

FIG. 9. Comparison by GLC of clove bud oils.

TABLE 3
Typical Applications and Inclusion Rate for CO_2OL Extracts in Foods

Juniper berry	100 ppm	Spirituous beverages; gin, liqueurs
Hop	10 ppm	Brewing (post-fermentation topnote); shandy
Ginger	0·1%	Beverages; ginger ale, ginger beer, liqueurs
Nutmeg	0·1%	Sauces, meat products
Vanilla	100 ppm	Natural baby foods and dairy products
Black pepper	0·1%	Meats, sauces
Clove bud	0·1%	Kretek cigarettes, toothpaste, chewing gum, sauces, apple products
Pimento berry	0·1%	Sauces, pickles
Cardamom	50 ppm	Liqueurs, health products
Mace	0·1%	Sauces, meat products; softer and less stringent than nutmeg

COMMERCIALLY AVAILABLE RANGE

Listed in Table 3 are product applications and typical use levels for some CO_2 extracts. These are best suited to, but not exclusively usable in, products which are not heat processed.

Terpeneless Oils

Truly terpeneless essential oils can be prepared from CO_2OL extracts by using the additional technique of extremely high vacuum molecular distillation with the minimum of heat. The process removes the 'absolute' character of the carbon dioxide extract to give an ingredient which is terpeneless. A terpeneless extract, in which the terpenes were never generated, is superior to conventional terpeneless oils in which the oil is steam-distilled from the botanical with heat and then heated again to fractionate off the terpenes. The cool method is much more refined and gives an ingredient of superior odour and solubility in which the terpenes were never formed.

Carbon dioxide extracts are cost-effective because they are often more concentrated than steam-distilled oils or extracts prepared with less selective solvents. They are also generally of finer odour and better fixed, hence longer lasting than steam-distilled oils.

In foodstuffs, not subjected to heating, a selective extract is a cleaner, altogether more elegant ingredient. Where many creations contain a preponderance of flavour chemicals, the benefit of using new natural products to give soul to a flavour skeleton is readily appreciated.

REFERENCES

1. Pickett, J. A., Coates, J. and Sharpe, F. R. (1970). *Chem. and Ind.*, 5th July, 571.
2. Moyler, D. A. (1984). *Perfum. Flavor.*, **9**, 109–15.
3. Gardner, D. S. J. G. (1982). *Chem. Ind.*, No. 12, 402.
4. Analytical Methods Committee of the Chemical Society, Essential Oils Subcommittee (1980). *Analyst*, **105**, 262–73.
5. Analytical Methods Committee of the Chemical Society, Essential Oils Subcommittee (1981). *Analyst*, **106**, 448–55.
6. Analytical Methods Committee of the Chemical Society, Essential Oils Subcommittee (1984). *Analyst*, **109**, 1339–41.
7. Analytical Methods Committee of the Chemical Society, Essential Oils Subcommittee (1984). *Analyst*, **109**, 1343–60.
8. Moyler, D. A. (1985). *Proc. Brit. Soc. Flavourists Symp.*, Paper No. 5.

9

β-Cyclodextrin and Its Uses in the Flavour Industry

James S. Pagington
Cairn Foods, Chesham, Bucks, UK

ABSTRACT

β-Cyclodextrin has been known for over 80 years. Originally thought to be poisonous, its complexing ability was considered a mere scientific curiosity, but extensive research in recent years has shown it to be not only non-toxic but a desirable food additive when used to protect flavours, vitamins and natural colours. Complexes produced using β-cyclodextrin resist oxidation, evaporation, degradation by UV and visible light, chemical attack and inter- or intra-molecular reactions. These products are not difficult to produce, and can be manufactured in a dry, crystalline form. Because of their stability, drying can often be carried out by the simple application of heat and/or vacuum. Many complexes are stable in solution as well as in the crystalline state, so it is possible to colour a soft drink with an insoluble, unstable natural colour. It is also possible to produce a stable onion or garlic oil spread, for coating onto a product, without loss of the oil prior to consumption. Equally, a benzaldehyde complex can be baked into a biscuit or put through an extruder without loss of flavour. β-Cyclodextrin can also be used to selectively extract components from complex mixtures. This opens the way for a variety of processes including the purification of allyl isothiocyanate, the removal of naringin from orange juice and the debittering of protein hydrolysates. In many such applications the β-cyclodextrin can be regenerated and recycled.

INTRODUCTION

History of Cyclodextrins

The first reference to cyclodextrins was by Villiers in an 1891 publication. It was not, however, until the 1930s that Freudenberg first prepared cyclodextrins. A detailed review of earlier works has been published by French.[1]

During the past ten years cyclodextrins have been investigated in laboratories around the world. Production plants now exist in both Hungary and Japan, and research into the manufacture of the product, especially in Hungary, has reduced the price by a factor of over 100.

The initial interest in cyclodextrins was for use with pharmaceutical products. With the reduction in price and increase in manufacturing capacity, that interest now includes their use in the chemical, agrochemical, cosmetic and food industries.

Manufacture of Cyclodextrins

Starch is a polymer of glucopyranose units with both 1–4 and 1–6 linkages. When treated with α-amylase the large polymeric molecules are split, and the smaller polymers commonly known as maltodextrins result.

Treatment of a maltodextrin solution (DE = 5) with the enzyme cyclodextrin transglycosylase (CTG) will link between 6 and 12 glucopyranose units head-to-tail in the 1–4 positions (Fig. 1). CTG is a very specific enzyme obtained by the cultivation of *Bacillus macerans*. To manufacture approximately 4 tons of cyclodextrin, the charge from a 1000-litre fermenter is needed. An excellent review of the work carried out on the CTG enzyme has been published by Szejtli.[2]

Chemistry of Cyclodextrins

A number of different cyclodextrins are formed by the above process, the most important being α, β and γ. These contain 6, 7 and 8 glucopyranose units respectively. Higher members of the series, containing 9, 10, 11 and 12 units and possibly side chains as well, have been identified. These products have negligible complexing ability, and thus are of little interest.

The main product is β-cyclodextrin (97%+) (Fig. 2). Purification of the α- and γ-cyclodextrins considerably increases their production

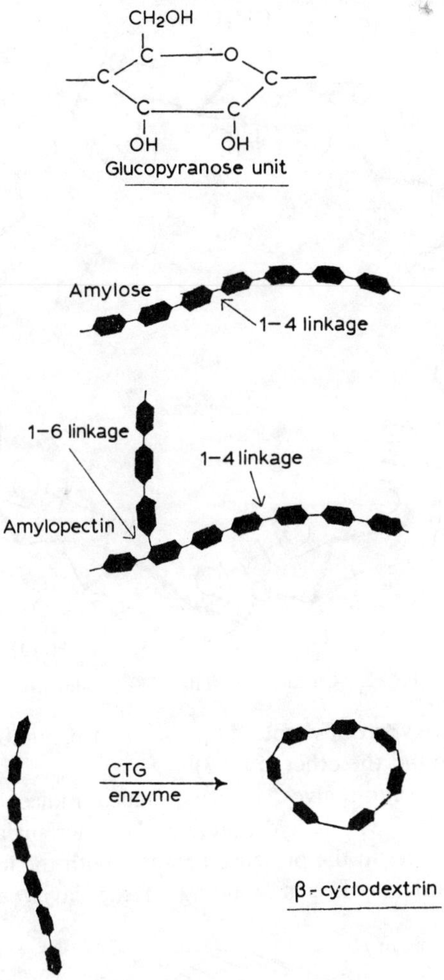

FIG. 1. Conversion of starch into cyclodextrins.

cost, and puts them into the fine-chemical classification, i.e. very expensive. This paper deals exclusively with β-cyclodextrin which is the only product commercially available in quantity.

The 7 glucopyranose units of β-cyclodextrin are linked in a cyclic structure at the 1–4 positions (Fig. 2). All the inter-ring oxygen bridges point inwards, making the centre of the molecule an area of high electron density. The molecular shape is a torus with the 14

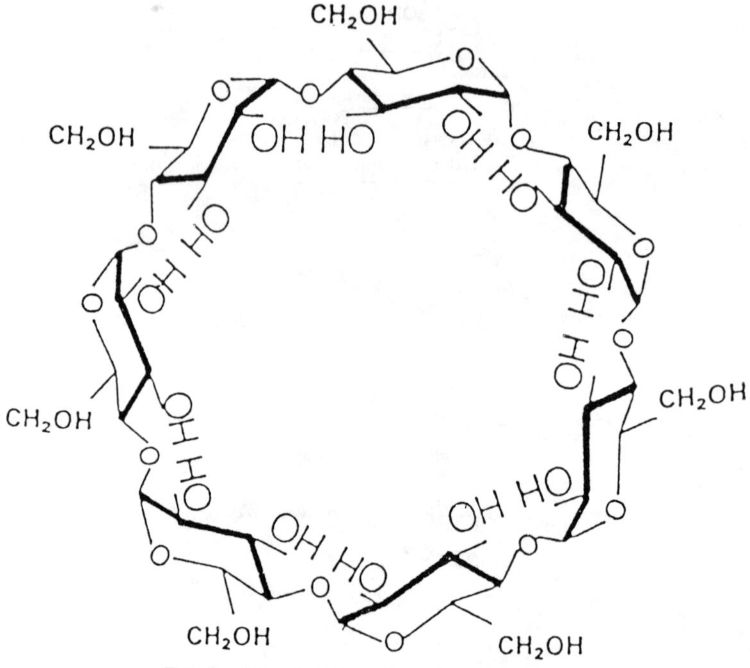

FIG. 2. Chemical formula of β-cyclodextrin.

primary hydroxyl groups around one outer edge, and the 7 secondary hydroxyls around the other (Fig. 3).

This configuration gives a torus shaped molecule; the cavity, measuring 7·8 × 7·8 Å, is hydrophobic, the outer surface being relatively hydrophilic. In the presence of water both the inner and outer surfaces hydrate; hydrogen bonding of the cavity water and the

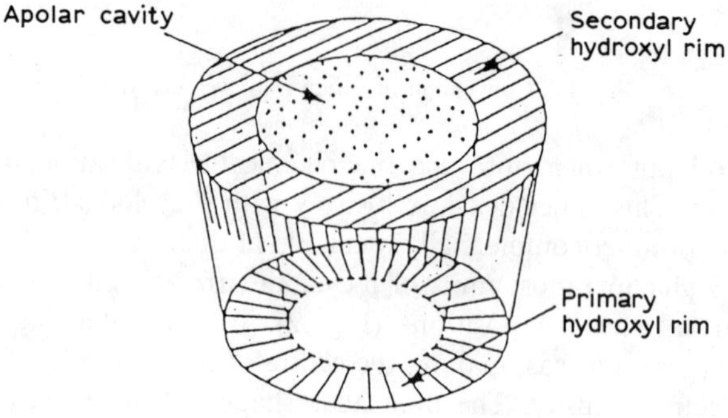

FIG. 3. Molecular shape of β-cyclodextrin.

TABLE 1
Chemical Specification of β-Cyclodextrin

Empirical formula	$[C_6H_{10}O_5]_7$
Molecular weight	1 135
Diameter of cavity	8 Å
Volume of cavity	346 Å^3
Water in cavity	11 molecules
Solubility in water at 25°C	0·85 g/100 ml
Purity (dry basis)	99·5 ± 2·5%
Other cyclodextrins	2·0% max. $\alpha + \gamma$
Moisture	14% max.
pH	6·5 ± 1·5
Optical rotation, $[\alpha]_D^{20°}$	+155° to +170°
Directly reducing matter	0·2% max.
Ash (sulphated, 550°C)	0·5% max.
Sulphur dioxide	10 mg/kg, max.
Protein (nitrogen)	N.D.

molecule itself causes distortion. The hydrated cyclodextrin molecule therefore is in a high energy state, and reduction of that energy by replacement of the cavity water is the secret of this remarkable product.

β-Cyclodextrin is a non-hygroscopic, white crystalline product with a slightly sweet taste. The Hungarian product assays at 100% cyclodextrin, with up to 2·5% of the α and γ products present. All three cyclodextrins are optically active. The properties of β-cyclodextrin are given in Table 1.

Cyclodextrin Derivatives

Fourteen of the 21 OH groups in the β-cyclodextrin molecule can be substituted in a variety of ways. The most interesting to date is methylation which produces a soluble derivative, heptakis-(2,6-*O*-dimethyl)-β-cyclodextrin. The reduced activity of the C-3 hydroxyls will not allow further substitution.[3]

Ether derivatives have also been produced containing allyl,[3] propyl,[4] ethyl and isopropyl benzenesulphonyl groups.[5] It is possible to produce ester and acid derivatives, also nitrogen and sulphur containing products.[6]

The OH groups of the β-cyclodextrin molecule can be chemically linked to each other or to some other material, such as methyl cellulose or cellulose acetate. This will produce a variety of polymers, including insoluble products where the complexing ability is retained.[7]

CYCLODEXTRIN COMPLEXES

Complex Formation

The ability of the cyclodextrin molecule to attract and hold other molecules in its cavity is the basis of its chemistry. In solution the hydrated cavity is an ideal environment for many hydrophobic molecules (Fig. 4).

There are three criteria which govern complex formation:

(1) The high energy state of the hydrated cyclodextrin molecule: this is energetically favourable for the molecule to replace the water with some other, more hydrophobic, guest.

(2) The hydrophobic nature of the guest molecule: the more hydrophobic the molecule, the more readily it will complex. Equally, when the aqueous system is in equilibrium, the most hydrophobic molecules will complex. The degree of dissociation of a complex, and therefore its stability in solution, is also related to the hydrophobic nature of the guest. As a guide, the insolubility of a guest molecule can be equated to its hydrophobic nature.

(3) The stereochemistry of the guest molecule: will all, part or just a side-chain fit into the torus? It is possible to obtain complexes with molar ratios of 1:2, 1:1 and 2:1, the most common being 1:1. The diameter of the β-cyclodextrin cavity is 7·8 Å; this can accommodate a naphthalene molecule.

The formation and breakdown of a complex must take place in solution, involving association/dissociation as shown in Fig. 4. If an insoluble material is to be complexed then it must be taken into

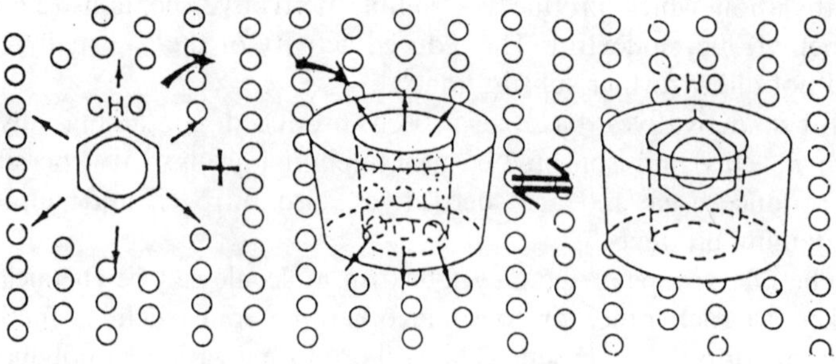

FIG. 4. Complexation of β-cyclodextrin.

solution using a water-soluble solvent. Once in solution, the most hydrophobic material will complex, i.e. the insoluble material will complex in preference to the soluble solvent.

Complex Manufacture

There are three basic methods of producing complexes:

(1) Co-crystallisation of the guest molecule and the β-cyclodextrin from solution (hot or cold, acid or alkaline). The solubility of β-cyclodextrin rises with temperature (Fig. 5). The easiest method of obtaining a complex is to crystallise it out by adding the guest to a hot solution of cyclodextrin, and then cooling. Insoluble compounds can be dissolved in a solvent. Products such as essential oils with limited solubility should be continuously agitated during cooling to encourage solubilisation.

(2) Mixing the β-cyclodextrin and guest in a blender with sufficient water to ensure a paste-like consistency. This may be a long process as sufficient time must be allowed for all of the cyclodextrin and guest molecules to dissolve in the small amount of water present. As only a small amount of each component is in solution at any time, equilibrium is not reached. The use of solvents in this process is inadvisable as they will complex, and tie up the cyclodextrin unnecessarily.

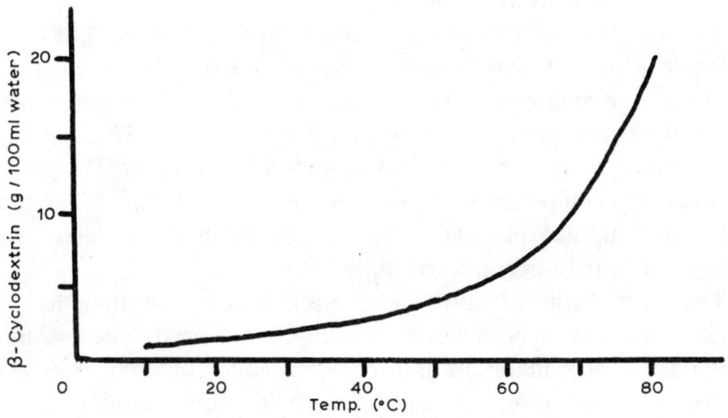

FIG. 5. Solubility curve of β-cyclodextrin.

(3) Passing the host in vapour form through a solution of β-cyclodextrin. The guest can be recovered either as a crystalline complex or in its original form by steam distillation of the complex solution or slurry.

In each of the above cases the crystallised material can be filtered and dried.

Drying may be by any conventional method including heating under reduced pressure or fluid bed drying. Losses of complexed volatiles are miminal as only the dissolved and dissociated molecules are lost.

When crystallised out of solution, up to 20% of the guest material may be trapped but not complexed. Such material will not be as stable as that which is complexed.

The percentage content of a complex will depend on the molecular weight of the guest material. β-Cyclodextrin has a molecular weight of 1134 and a 1:1 complex with a typical essential oil (molecular weight range 90–130) will contain 8–12% essential oil.

Complexes tend to be less soluble than the uncomplexed β-cyclodextrin; even so, complexes of very insoluble materials can often improve the guest's solubility by several orders of magnitude.

Complex Stability

Powdered complexes

Complexation will protect a guest molecule from loss by evaporation and from attack by oxygen, UV and visible light, as well as inter- or intra-molecular reactions.

The resistance of a flavour complex to attack by oxygen has been demonstrated with the β-cyclodextrin complexes of benzaldehyde, garlic oil and lemon oil.[8] These materials and their complexes were exposed to neat oxygen, in a Warburg apparatus, at 37°C. In each case the oxygen uptake of the flavour materials was rapid. The oxygen uptake of the complexes was both slow and limited. Figure 6 illustrates the limited uptake of oxygen by the benzaldehyde complex when compared to the uncomplexed material.

The effect of complexation on flavour loss by evaporation has been demonstrated with both liquid (anethole) and crystalline (vanillin) materials.[8] These materials and their β-cyclodextrin complexes were stored as 2 mm layers in open vessels for the duration of the experiment. The free anethole had disappeared by day 16, whereas

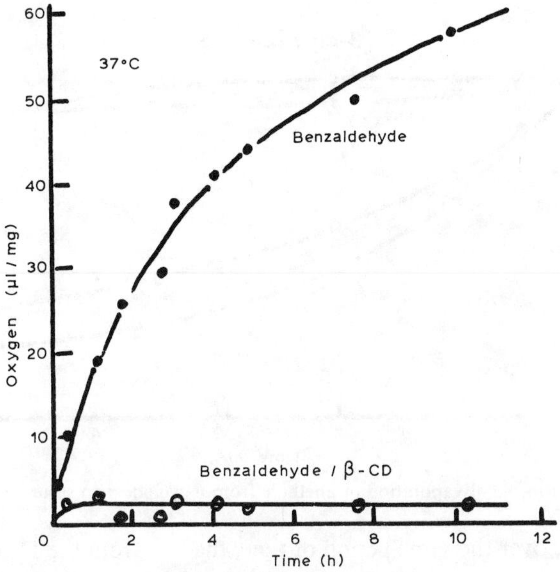

FIG. 6. Oxygen uptake of benzaldehyde and its β-cyclodextrin complex.

less than 5% of the complexed anethole had been lost. All of the vanillin had been lost by day 240, whilst only 19% of the complexed vanillin had evaporated (Fig. 7). It has subsequently been reported that no further losses of vanillin were recorded even after 1000 days.

The stability of vitamin D_3 and its β-cyclodextrin complex has been studied,[9] both in the presence of copper sulphate and by irradiation with light. In both cases the free vitamin decomposed

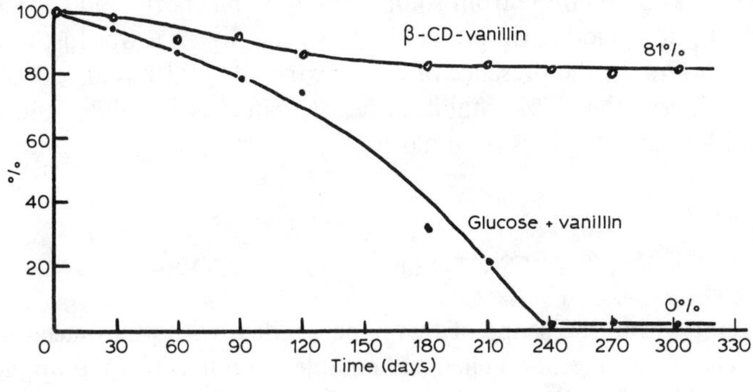

FIG. 7. Evaporation of vanillin and its β-cyclodextrin complex.

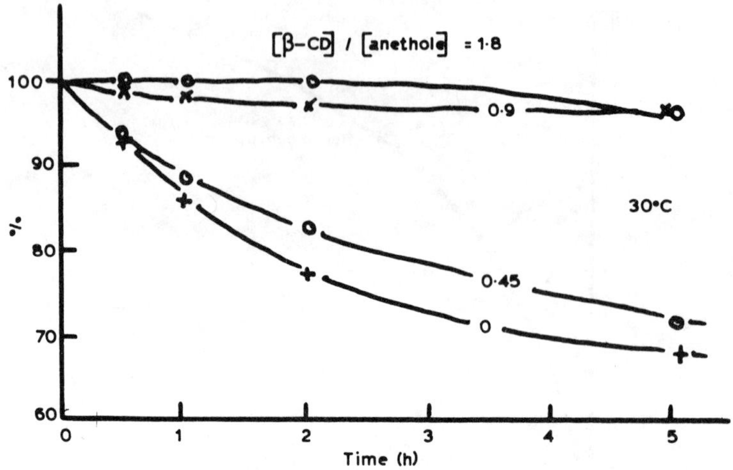

FIG. 8. Evaporation of anethole from β-cyclodextrin solution.

whereas, over the same period of time, the loss from the D_3 complex was minimal.

Loss of a guest material from its complex can often be associated with trapped, but uncomplexed, material. This can vary from 1% to 20%, depending on the manufacturing procedure and to some extent the material itself.

Stability in solution

Active and labile substances can be stabilised in solution. The degree of stabilisation will depend on the degree of dissociation, i.e. how much free material is present.

The loss of anethole from solution at 30°C has been studied[8] (Fig. 8). With no cyclodextrin present, 68% of the material was lost in the first 5 hours. In the presence of cyclodextrin at a 1:1 molar ratio, the loss was less than 5%. Stabilisation of materials in dilute solution could be advantageous in liquid food products, fruit juices and soft drinks.

TYPES OF FOOD COMPLEXES

It is not within the scope of this paper to deal with pharmaceutical, agrochemical or general chemical complexes, nor with food applications other than those applicable to flavours. It is, however, worth making a brief reference to several classes of compounds often used alongside flavours, and of interest to the flavourist.

Colour Complexes

The majority of natural colours are unstable and of limited solubility. These factors restrict their application, despite the current pressure to find stable replacements for the synthetic colours.

Many fat-soluble natural colours can complex with β-cyclodextrin. Complexes have been reportedly prepared with carotenoids, paprika, turmeric and annatto.

Carotenoids are used at levels of 1–9 mg/litre in soft drinks and up to 30 mg/kg in confectionery products (Table 2).[10] Complexes of these colours, whilst having a limited solubility, should be sufficiently soluble to colour the final products.

Bixin, when dissolved in the presence of β-cyclodextrin, did not fade on exposure to light, even in a clear glass container.

Loss of colour on processing and storage can be considerably reduced by the addition of β-cyclodextrin. In some instances, e.g. tomato ketchup, addition of 0·2% to the product is all that is required to stabilise the colour, even when the product is processed at 100°C for 2 hours.[11]

Vitamin Complexes

Vitamin/β-cyclodextrin complexes, like those of other products, are stable to heat, oxidation, chemical attack, etc. When vitamins are added in complex form, they no longer need to be overdosed, this represents considerable savings. They are protected during processing and storage.

TABLE 2
Levels of Carotenoids Required to Colour Some Beverage and Confectionery Products[10]

Product	*Carotenoid level (pure substance) (mg per litre or kg)*
Fruit juice beverages	2·5–5·0
Fruit syrups	1·0–8·0
Carbonated beverages	1·0–1·5
Hard boiled sweets	2·5–30
Pectin jellies	3·9–30

Extensive work on the stability of vitamins A, B_3, E and K_3 has been carried out.[9,12] Complexation has been found to increase both solubility and bioavailability.

Szejtli has shown that the solubility of vitamin D_3 (at 37°C and pH 7) increased from less than 1 μg/100 ml to over 35 μg/100 ml in the presence of β-cyclodextrin.[9]

Sweetener Complexes

Dihydrochalcone type sweetening agents have been stabilised in solution by β-cyclodextrin. The addition of cyclodextrin to orange juice has prevented precipitation of the sweetener at low temperature.[13]

It is possible to use cyclodextrin with a number of sweeteners to eliminate aftertaste where this is a problem. A mixture of aspartame and β-cyclodextrin is marketed in Japan under the name Pal Sweet. Little information is available to date regarding the actual stabilising of sweeteners in dilute solution.

Flavour Complexes

Many flavouring materials are volatile liquids, often subject to oxidation. The use of such materials, especially when mixed in a dilute powder form, presents many problems. They can be spray-dried onto a gum, starch or maltodextrin base, or spread onto a salt, sugar or starch base. Both methods have their associated problems:

(1) Spray drying is an expensive process. Volatile materials are often lost, and must be overdosed to compensate. Contamination is a serious problem and necessitates extensive cleaning of equipment. This restricts the manufacturing time considerably, especially if the programme includes a large number of small volume items. Strength of a spray-dried flavour can often be a problem. The quantity of flavouring material which can be dried onto a base is limited.

(2) When spread onto a base such as salt, flavouring materials are exceptionally vulnerable as they are dispersed over a large surface area. Under such conditions materials are rapidly lost due to evaporation and/or oxidation. When the flavouring material is a liquid, the level of addition is restricted if the final product is to be free-flowing. Most of the bases in common use are hygroscopic and must be protected from the atmosphere.

All the foregoing problems can be solved by the use of β-cyclodextrin. Now that the price has fallen it can, in combination with maltodextrin, compete with other spray-drying carriers. Typically, approximately 10% of a solvent-free flavour may be dried onto cyclodextrin and then diluted to a specified strength. Alternatively, products such as fruit juices can be dried onto a suitable base with addition of sufficient cyclodextrin to complex the volatiles present.

Many cyclodextrin complexes do not require spray-drying. They can be produced by crystallisation, and dried by cheaper methods as described earlier in this paper.

When using β-cyclodextrin the flavourist should re-examine his product. If it is uneconomic to complex a whole flavour, even after removal of the solvents, then individual components should be

TABLE 3
Flavour Content of Some β-Cyclodextrin Complexes

Flavour material	*Flavour content of complex (%)*
Anise oil	9·00
Basil oil	10·72
Bay leaf oil	10·80
Benzaldehyde	8·70
Caraway oil	10·50
Carrot oil	8·82
Celery oil	10·00
Cinnamon oil	8·76
Coriander oil	7·72
Dill oil	6·92
Fume oil	12·20
Garlic oil	10·20
Lemon oil	8·75
Lovage oil	4·35
Marjoram oil	8·00
Mustard oil	10·92
Onion oil	10·20
Orange oil	9·20
Peppermint oil	9·70
Raspberry oil	8·66
Sage oil	8·20
Sweet cumin oil	10·00
Tarragon oil	10·23
Thyme oil	9·60
Vanillin	6·20

considered. The problem ingredients can be complexed and subsequently blended into the dry bulk of the flavour. This could include such items as DMS, acetaldehyde, pyrazines, sulphur-containing compounds, and the oils of onion and garlic. Such a process will avoid expensive losses during both processing and storage. Deterioration of flavour in the finished food can also be prevented.

A list of selected flavouring complexes is given in Table 3. It is important that, when a flavour is released in the mouth, its composition is unchanged from that of the original. This has been thoroughly investigated chromatographically[8] using the oils of dill, onion and marjoram. In each case the recovered material was of similar composition to the original oil (Fig. 9).

When an excess of potential guest molecules are present in solution, the most hydrophobic ones will complex first. This can be useful in the purification of a material, or in the removal of an impurity.

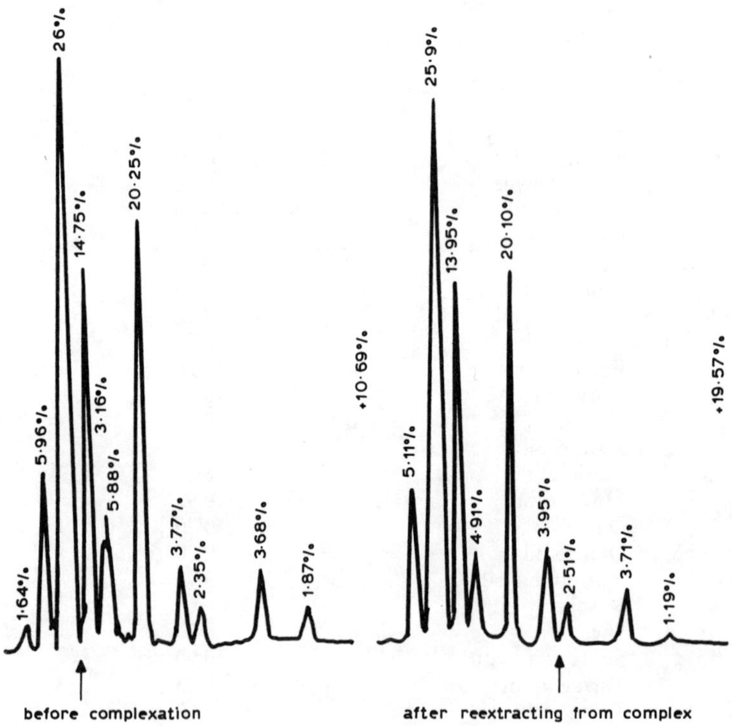

FIG. 9. Chromatograms of marjoram oil before and after complexation.

Equally, when a complex is produced with excess of a multicomponent flavouring, the complex may not accurately represent the original. For example, an excess of lemon oil would result in formation of a terpene complex, with the important oxygenated flavour compounds left in solution.

To date, β-cyclodextrin has been viewed as an expensive ingredient which has to undergo processing to produce complexes. Currently it is being looked at as an additive to food systems, where it will complex materials in solution to give a desired effect.

Cyclodextrin will also function in the protection of noxious and labile materials. I hope that it will be possible to purchase many flavour components, not as a 1% or 5% solution, but as a 5% or 10% complex. Such complexes require no special packaging, transport or handling.

APPLICATIONS OF CYCLODEXTRINS

Food Products

Soft drinks

In very dilute solution, complexes of all but the most insoluble materials would be expected to dissociate. If any degree of dissociation exists, then the stabilising effects of complexation will not protect the free guest molecules from attack. It is unlikely therefore that cyclodextrin will protect water-soluble sweeteners. The limited amount of published information on sweeteners relates to masking or removing aftertaste rather than protection.

Complexes of insoluble colours are the most promising for soft drink application. Complexation should improve solubility and prevent fading. It should, however, be noted that, where the colour complex may be sufficiently soluble to colour a soft drink, it is unlikely to be sufficiently soluble for use in a concentrate.

Biscuit and confectionery products

The stability of β-cyclodextrin complexes at elevated temperature offers an ideal way of protecting volatile materials during processing. The coating of a complex with a hardened vegetable oil will give a very heat-stable flavouring.[14]

Complexes of flavours such as benzaldehyde or lemon oil can be added to a biscuit dough and baked without significant loss, the flavour only being released when the product is eaten. Flavour stability may be so effective that it is necessary to dust the product with a conventional flavour. This will give an aroma and initial impact to the food whilst the complex will give a long-lasting flavour in the body of the product.

The use of a peppermint complex in chewing gum will give a long-lasting flavour; it should also be easier to blend than the liquid oil. Both colours and flavours can be protected when used in all kinds of confectionery.

The non-hygroscopic nature of cyclodextrin can be useful in the preparation of dusting powders. A mixture of 5% soybean oil complex and 95% sugar protects sugar confections at relative humidities up to 80%.[15]

Snack food products

Colour, vitamin and flavour complexes can be used in an extruder without significant loss. The stability of these complexes should prevent the need for overdosing, and increase the variety of flavouring materials available for this application.

It is impossible to use a number of ingredients for dusting onto snack products such as potato crisps. Flavourings such as onion oil would be lost from the crisp during the product's shelf life. If applied as a complex, however, the volatile components are well protected, until they are being eaten. At that point the complex dissolves in the warm moisture of the mouth, dissociates, and the guest material is released and tasted.

Flavour/odour reduction in food products

The addition of cyclodextrin to a number of products can reduce unwanted and bitter flavours.[16] In the case of some drugs the unwanted odour of the product can also be removed.

The presence of free guest molecules in a product will depend on temperature, concentration of components and the complex dissociation constant. Thus the addition of low levels of cyclodextrin to warm tea or coffee is unlikely to complex unwanted flavours. The use of cyclodextrin in cold, concentrated solutions, however, will reduce flavours and odours by complexation.

The bitter off-taste, often present when protein hydrolysates are produced, can be eliminated by the addition of cyclodextrin.[17]

Tobacco Products

Aroma substances when blended with tobacco are easily lost during processing and storage. If they are applied as cyclodextrin complexes these losses can be avoided. Such complexes can either be dusted or sprayed as an aqueous suspension onto the tobacco, and the aroma will not be liberated until the tobacco is burnt.[18,19]

Cyclodextrins, in both monomeric and polymeric form, have been used successfully in cigarette filters.[20–22] Typically, a cellulose acetate filter was impregnated with 80·5 mg of β-cyclodextrin. At a smoking rate of 200 ml/min, the filter retained 68·5 mg of tar and 2 mg of nicotine. This compared with a retention of 13·2 mg of tar and 0·9 mg of nicotine with the untreated filter.

Cosmetics and Personal Hygiene Products

The ability of a cyclodextrin complex to release its guest molecule on contact with warm moisture makes it an ideal carrier for use in body powders. Whether the guest is a perfume, a drug or a deodorant, it will remain protected until released by perspiration.

A β-cyclodextrin/iodine complex has successfully been used as a deodorant and antiseptic,[24] and an ethanol/β-cyclodextrin solution has been recommended as a deodorant.[25] Many other possible applications have been researched,[26] including fragrant candles, solid perfumes, bath preparations and perfume complexes in washing powder.

Processing Applications

Selective complexing

In the presence of excess guest material, β-cyclodextrin will complex with the most hydrophobic molecules present. It is therefore possible to selectively remove the terpenes from an essential oil, to remove naringin from orange juice, or to concentrate allyl isothiocyanate from impure mustard oil[27] (Fig. 10).

Cyclodextrin molecules are optically active and have been used to separate the optical isomers of a number of compounds, both analytically and as a process method.

Concentration of flavouring materials

The treatment of dilute aqueous solutions of some materials with cyclodextrin can often result in their total or selective removal. Orange juice aroma can be removed and thus concentrated by extraction

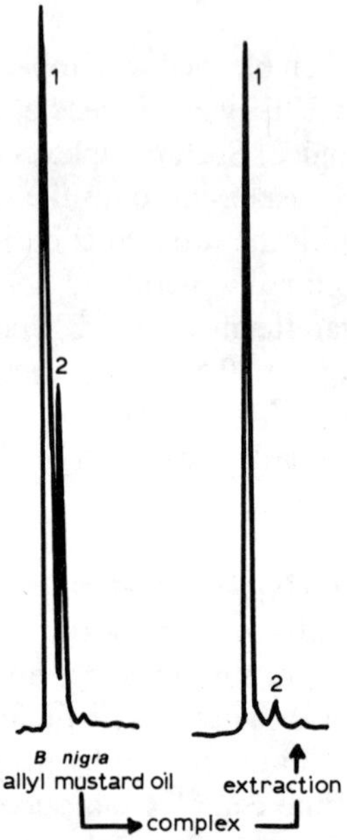

FIG. 10. Chromatograms from purification of allyl isothiocyanate.

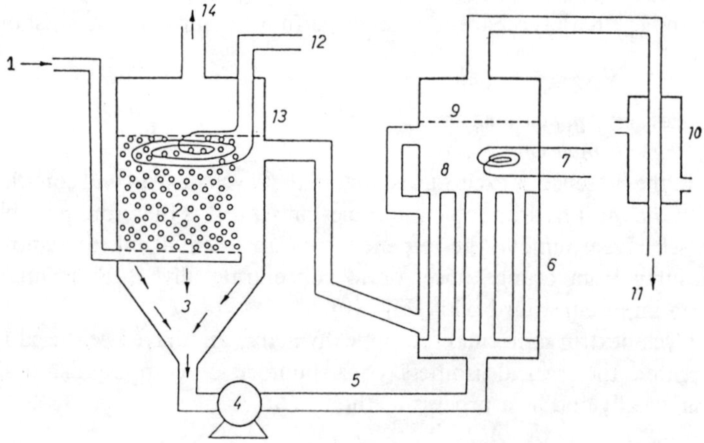

FIG. 11. Removal of volatiles from the vapour phase with β-cyclodextrin: 1, vapour phase of large volume; 2, saturated cyclodextrin solution; 3, settled crystals; 4, pump; 5, cold slurry; 6, heat exchanger; 7, supplementary calorifer; 8, warm solution; 9, solvent vapours; 10, condenser; 11, condensed solvent; 12, supplementary condenser; 13, cold cyclodextrin solution; 14, exit, scrubbed gas.

with β-cyclodextrin.[28] Flavour materials can be extracted from a complex mixture of vegetable or animal materials, and valuable materials recovered from effluent streams.

Flavour and solvent materials can be removed from the vapour phase by scrubbing with a β-cyclodextrin solution.[29] The guest material can be recovered by steam distillation and the cyclodextrin solution recycled to the scrubber (Fig. 11).

Chemical Reactions using Cyclodextrin

A number of stereoselective reactions can be carried out based on β-cyclodextrin inclusion complexes.[30] The cyclodextrin molecule can also act as a catalyst in many reactions.[31]

THE LEGAL STATUS OF β-CYCLODEXTRIN

Cyclodextrins have been accepted for use in food products in both Japan and Hungary for a number of years. During the last eighteen months applications have been made in France and the USA for its acceptance as a food ingredient.

In the UK, cyclodextrins fall into the FACC definition of a maltodextrin.[32] When asked for confirmation, the Committee ruled that β-cyclodextrin should be considered as a modified starch and not a maltodextrin. This will bring the product within the scope of their proposed regulations.

We understand that at present β-cyclodextrin is permitted for use in food under current law, but will not be permitted when the Committee's recommendations eventually become law. (This is waiting for the EEC regulations on chemically modified starches.) We have accordingly applied for β-cyclodextrin to be added to the proposed list of approved modified starches. The Food Advisory Committee and the Committee on Toxicology are currently considering the matter. All relevant toxicological data on both cyclodextrin and the CTG enzyme have been submitted, and we can only wait patiently for their judgement.

ACKNOWLEDGEMENTS

I would like to thank Dr J. Szejtli for his help, and for his permission to reproduce the diagrams used in this paper.

REFERENCES

1. French, D. (1957). *Adv. Carbohydr. Chem.*, **12**, 189.
2. Szejtli, J. (1982). *Cyclodextrins and Their Inclusion Complexes*, Akademiai Kiado, Budapest, 59–73.
3. Szejtli, J. (1982). Ref. 2, 78.
4. Bergeron, A. R., Machida, Y. and Bloch, K. (1975). *J. Biol. Chem.*, **250**, 1223.
5. Kurita, H., Kawazu, M. and Takashima, K. (1974). Jap. Kokai 74,85,014.
6. Szejtli, J. (1982). Ref. 2, 82–4.
7. Zsadon, B. and Fenyvesi, E. (1981). *Proc. 1st Int. Symp. Cyclodextrins*, Ed. Szejtli, J., Akademiai Kiado, Budapest, 327–36.
8. Szejtli, J., Szente, L. and Banky-Elod, E. (1978). *Acta Chim. Acad. Sci. Hung.*, **101**, 27–47.
9. Szejtli, J., Bolla-Pusztai, Eva, Szabo, P. and Ferenczy, T. (1980). *Pharmazie*, **35**, H.12.
10. Counsell, J. N. (1975). *IFST Proc.*, **18**(3), 156.
11. Kawashima, F. (1979). Jap. Kokai 79,117,536.
12. Szejtli, J. (1982). Ref. 2, 239–43.
13. Hashimoto, S. and Katayama, A. (1975). Jap. Kokai 75,35,941.
14. Ogawa Co. Ltd. (1981). Jap. Kokai 81,92,754/5.
15. Nawata, Y., Yamamoto, K. and Sano, M. (1977). Jap. Kokai 77,12,684.
16. Hamilton, R. M. and Haedy, R. E. (1970). US Pat. 3,528,819.
17. Suzuki, Y. and Ikura, H. (1975). Jap. Kokai 75,58,208.
18. Bavley, A. and Robb, E. W. (1962). US Pat. 3,047,431.
19. Robb, A. W., Westbrook, J. J. and Bavley, A. (1964). *Tobacco Sci.*, **8**, 3.
20. Arakawa, K. (1962). Ger. Offen. 2,527,234.
21. Miwa, K. and Tanata, Y. (1975). Jap. Kokai 75,125,100.
22. Miwa, K. and Tanata, Y. (1976). Jap. Kokai 76,32,799.
23. Buckler, S. A., Martel, R. F. and Moshy, R. J. (1964). US Pat. 3,472,835.
24. Hirose, T. (1976). Jap. Kokai 76,88,625.
25. Watanabe, Y., Ishizone, H. and Kawai, N. (1978). Jap. Kokai 78,41,440.
26. Szejtli, J. (1982). *Starch Starke*, **34**(11), 379–85.
27. Szejtli, J. (1982). Ref. 2, 259–60.
28. Buckler, S. A., Martel, R. F. and Moshy, R. J. (1969). US Pat. 3,472,835.
29. Szejtli, J., Bolla, E., Kormoezy, Gy., Nagy, G. and Nagy, Zs. (1978). Hung. Pat. **19**, 951.
30. Szejtli, J. (1982). Ref. 2, 269–72.
31. Szejtli, J. (1982). Ref. 2, 144–62.
32. Food Additives and Contaminants Committee (1980). *Report on Modified Starches*, FAC/REP/31, HMSO.

10

Flavour of Dairy Products

J. ADDA

Institut National de la Recherche Agronomique, Jouy-en-Josas, France

ABSTRACT

The flavour of dairy products has been extensively studied. Despite many results, which give a satisfactory understanding of the mechanisms involved, we are often not yet able to differentiate analytically between samples which are perceived as different on tasting, nor are we able to control the formation of the flavour. The role of lactic acid itself raises some questions as the acidification step seems to determine the final flavour quality of ripened cheeses. Besides lactic acid, the lactic flora produces acetaldehyde and diacetyl which play major roles in some products. Among compounds arising from fat degradation, free fatty acids appear to be important as precursors of flavour compounds such as methyl ketones, essential for Blue cheese characteristics or as precursors of compounds arising from oxidation, which may have a negative or a positive effect. One example of this, 1-octen-3-ol, which is objectionable in butter, is of importance for the flavour of mould-ripened cheeses. Whatever the importance of these, the main route to flavour formation, at least in cheese, is protein degradation, which leads to essential volatile compounds such as sulphur compounds but also to non-volatile compounds which have been much less studied. One of the main questions which is yet to be answered, is how can initial conditions later have so profound an influence on the evolution of the protein fraction.

INTRODUCTION

The flavour of dairy products originates from microbial, enzymic and chemical transformations, the relative importance of which is not always fully understood. These transformations give rise to a series of volatile and non-volatile compounds, some of which have been shown to correlate well with some typical flavour notes or flavour defects. However, very often compounds isolated from dairy products have been listed among flavour compounds with very little indication of their real contribution to perceived sensation. We will try, in this paper, to point out what available information has been already discussed in various reviews[1-5] and is to be considered as relevant for further studies.

COMPOUNDS ORIGINATING FROM LACTOSE AND CITRATE

The lactic flora activity leads to a series of compounds which are of importance to most dairy products.

Lactic Acid

Lactic acid itself is of special importance, even if not often mentioned by flavour chemists, primarily for the acidity it creates in the various cultured products, but also because the pH drop, which follows lactic acid production, determines the basic structure, characteristic of each cheese variety. The mineral content of the curd also seems to be involved in a definite but unexplained relationship between cheese texture and flavour since a poorly textured cheese almost always develops an unsatisfactory flavour.[6,7] The pH drop also governs the proportions of residual chymosin and plasmin in the cheese. Lactic acid may also lower the proteolytic and lipolytic activities at least in soft-ripened cheeses,[8] and lactic acid has thus an indirect but essential influence on the degradation of the main cheese constituents, which ultimately will influence the formation of flavour compounds.

Lactic fermentation has also been held responsible for the reduced environment (negative oxidation/reduction potential) which prevails in cheese. This reduced environment has been thought to be, in Cheddar but also in Swiss cheese, the main condition for flavour development.[9-11]

Another role that can be assigned to lactic acid is to serve as a substrate for various microorganisms such as moulds and propionibacteria. In this latter case lactate will lead to propionate, the importance of which is known for Emmental and Gruyère where it has been held responsible for the sweet taste usually found.[12]

Acetaldehyde

The next metabolite of importance is acetaldehyde. It is recognised as giving the characteristic flavour to yoghurt which is obtained by the growth of mixed cultures of *Streptococcus thermophilus* and *Lactobacillus bulgaricus*, but it also appears as a by-product in cultured cream and butter as a metabolite of *Str. cremoris diacetylactis* and *Leuconostoc citrovorum.*[13]

In yoghurt, the acetaldehyde level differs greatly with different strains,[14] and there is no agreement on the acetaldehyde values to be obtained for an optimum flavour. For some the optimum lies in the range 23–41 ppm while, for others, yoghurt of good flavour quality contains 13–16 ppm, and still lower values can be encountered in commercial products.[2,15] Furthermore, little is known of the influence of other volatile compounds such as 2-butanone and diacetyl on aroma and of peptides on taste. Yoghurt quality is greatly influenced by its texture and viscosity, and more research is necessary to understand their effects on the perception of flavour. The observation that yoghurt made from ewe milk has a milder and richer flavour than yoghurt made from cow milk may perhaps be explained by its higher protein content.

Diacetyl

Diacetyl is an important flavour constituent of cultured products. It can be formed from the citrate present in milk by *Str. diacetylactis*, *Leuconostoc citrovorum* and other organisms that can utilise citrate.[16]

Diacetyl strongly contributes to acid cream butter flavour, where it is usually found at the 1–2 ppm level,[13] but a too high level is known to lead to flavour defects. Many other minor components, including 2,3-pentanedione, have been additionally found in acid cream butter[17] but their sensory contribution is not yet fully understood.

Diacetyl plays a major role in unripened soft cheeses such as cottage cheese where a desirable level is around 12 ppm.[18] It has also been considered as important to Swiss cheese flavour[19] where it has been correlated with the sweet note that can be perceived in the water-soluble volatile fraction.[20]

The possibility to enhance the diacetyl level in cheese by encapsulating bacterial cell-free extract has been demonstrated.[21] An eightfold increase of diacetyl was thus obtained but unfortunately the authors made no comment on the resulting flavour.

Diacetyl is easily reduced to acetoin[16] which is also present at relatively high level in cultured products.[13] This compound can be further reduced to butylene glycol, which can lead to 2-butanone as a result of *Lactobacillus plantarum* and *L. brevis* activity. Such a mechanism has been demonstrated in Cheddar.[22]

THE ROLE OF FAT IN THE FLAVOUR

The Importance of Milk Fat

The role of milk fat is recognised as important for different sensory attributes of dairy products. It is a common observation that too high or too low a fat level leads to poor quality. It is difficult to decide if the reason for this is linked to differences in texture or viscosity, to differences in the amount of volatiles arising from milk fat, or if the main reason is that the fat phase which serves as a solvent for flavour compounds then interacts differently.

Some experiments have shown that milk fat is important in dairy products because of some intrinsic qualities. For example,[23] it has been shown that Cheddar cheese made with vegetable or mineral fat does not develop the right flavour and, moreover, the quality of the cheese is not optimal if the milk fat is structurally modified by mechanical action before being reincorporated in the milk. This suggests that the nature of the oil/water interface which exists in the milk fat globule may be of importance, as could be the physical characteristics of the fat to allow an optimum perception of flavour compounds.

Free Fatty Acids

The role of milk fat as a precursor of flavour compounds is also of importance. Triglycerides account for about 98% of milk fat. After the action of native or exogenous enzymes, alkanoic acids are set free, sometimes in very large amounts. The importance of free fatty acids (FFA), which differ considerably in chain length and flavour intensity, varies between products, but as a whole FFA are more often a source of problems than a factor of quality.

This especially applies to cow milk where the presence of even a

TABLE 1
Rancidity Score in Butter[28]

$$\text{Rancidity score} = 1{\cdot}931\,63 + 0{\cdot}017\,54\ \text{butyric acid} + 0{\cdot}000\,54\ C_{18}\ \text{FFA congeners}$$

small amount of short chain FFA leads to an easily perceived rancid flavour. At the opposite extreme in goat milk, FFA have for long been held responsible for the specific flavour[24] which gives goat milk an important premium. However, when looking closer at the problem it appears that the role of FFA has perhaps been overestimated[25] and FFA are probably only one of several factors causing 'goat flavour'. The recently published results according to which 4-ethyloct-2-enoic acid could play a major role[26] need further confirmation before we are able to explain the concern which is sometimes expressed about a real or imagined disappearance of the intensity of goat flavour in herd milk.

The role of FFA in butter flavour does not seem very important, except perhaps for C_{10} and C_{12} alkanoic acids which have been shown to be present in sweet cream butter at levels exceeding their flavour threshold.[27] On the other hand, FFA (and mainly short chain ones) are responsible for hydrolytic rancidity which is still an important problem for the butter industry. It may develop during frozen storage, as a result of microbial lipase activity, which has been shown to resist even severe heat treatment. An important point is to be able to detect early deterioration and even to predict forthcoming rancidity. The existing methods, based on titration of total FFA, do not permit easy detection in nearly rancid samples. Correlating sensory data with individual FFA concentration by the use of stepwise regression and discriminatory analysis brings an elegant method to index rancid flavour intensities in butter.[28] A rancidity score is thus obtained (Table 1) which allows us to classify butter samples (Table 2).

TABLE 2
Classification of Butter Samples[28]

Grouping	*Range of rancidity scores*
Non-rancid	1·00–2·80
Slightly rancid	2·81–3·80
Rancid	3·81–7·00

High levels of FFA are to be found in most types of cheese. However, there do not seem to be any correlations between FFA and flavour intensity except perhaps in Blue cheese where the observed levels are above the flavour thresholds. Rancidity is nevertheless observed in some cases, either when too high a level of butyric acid is attained as a result of *Clostridium tyrobutyricum* activity, or as a result of a too high lipase activity. This can be observed in various types of cheese made from milk contaminated with psychrophilic microflora or when lipase has been added with the aim of accelerating ripening as observed in Cheddar,[29] for example, or when natural lipases have been activated by a process such as milk fat homogenisation.

Besides a direct negative influence on flavour, too high a FFA level may have an adverse effect on cheese quality by inhibiting some microorganisms responsible for the biosynthesis of other important flavour constituents.

The practice of adding kid lipase, which preferentially hydrolyses lower chain fatty acids, brings good results for certain Italian types of cheese such as Romano, Provolone and Fontina where rancidity does not appear as a defect,[30,31] and some positive results have been claimed for Cheddar after the action of added fungal lipase.[32]

Hydroxyacids and Lactones

Besides alkanoic acids, milk fat contains small amounts of δ-hydroxyacids which in certain circumstances spontaneously lactonise to form the corresponding δ-lactones.

Unsaturated C_{18} fatty acids may lead, via hydration, to hydroxy-acids which, via β-oxidation, lead to γ-lactones;[17] these also occur in butterfat. The amount of lactone precursors present in milk varies with feed, season, breed and stage of lactation.[33] Supplementing the diet with fat reduces the level of δ-lactone precursors while the amount of γ-dodecalactone precursors can be increased by including a readily available source of starch in the diet.[34]

Lactones have been considered as important contributors to butter flavour. Their real importance is difficult to assess since large discrepancies are to be found between figures: some authors[35] indicate values of 7 ppm in freshly made butter, compared to values of 85 ppm given by others.[36] The large observed variation between samples of the same quality led Stark and Urbach[37] to the conclusion that there is no relationship between the level of free lactones and

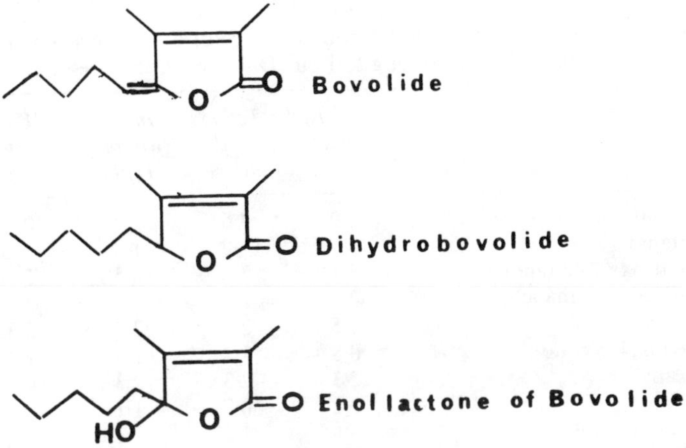

FIG. 1. Some unsaturated lactones of milk fat.

the quality of butter even if δ-octa- and deca-lactone seem to participate in the basic butter flavour, as a consequence of being present above their flavour threshold.

When milk fat is heated the lactone levels increase, and some unusual unsaturated lactones (Fig. 1) have been reported in heated butter.[34] Heat treatment of milk results also in the formation of lactones. The influence of heating on milk is well documented. Initial work has dealt with sterilised milk, but the growing importance of the UHT process has diminished the economical importance of the former. The acceptability of UHT milk was initially low for those people used to pasteurised milk, so UHT off-flavour has been carefully examined.[2,38,39]

The commercial success of UHT milk, at least on the Continent, makes it difficult to keep talking of an off-flavour, and even perhaps to keep pasteurised milk as a reference. Nevertheless, some people are still concerned about UHT milk flavour and are trying to obtain a product with the same keeping quality but closer in flavour to pasteurised milk.

Table 3 shows the preliminary results obtained on analysing the major volatile compounds extracted from four samples all made from the same raw milk. Sample 1 is a low temperature pasteurised milk; sample 2 is obtained with modified Laguilharre equipment[40] (see Fig. 2) in which a Teflon insert avoids any contact of milk with heated metal surfaces; sample 3 is obtained with the same equipment

TABLE 3
Amounts of Different Volatile Compounds (μg/kg) Extracted from Heat-treated Milk According to Four Different Procedures

	Pasteurised milk	*Teflon lined insert*	*Insert without Teflon*	*UHT-i*
2-Pentanone	0·9	4·5	5·5	30
Pentanal	3	1·3	1·3	3
3-Hydroxy-2-butanone	3·1	3·6	0·5	5
1,1-Diethoxyethane	2	2	2	3
Pyridine	2·5	2·2	2·7	3·3
Ethyl isobutyrate	0·8	1·2	1·3	1·4
Hexanal	2·3	2·3	2·3	4
Furfural	0·2	0·7	1·1	2
Hexanol	3	1·6	2·2	4
2-Heptanone	0·9	19	21	83
Heptanal	0·7	0·7	0·7	2·1
Benzaldehyde	1·2	5·5	5·3	5
Octanal	0·7	0·7	0·6	2
Hydrocarbons	—	—	—	+ + +
Acetophenone	2	1·7	2·1	2·6
2-Nonanone	0·9	16	15	45
Undecane	—	6	4·6	
Nonanal	2·5	2·3	2·1	5
Hydrocarbon	0·5	4·1	3·3	
Decanal	0·8	1	0·8	1·6
Benzothiazole	4·5	4·2	5·5	6
Unknown	1	3·9	0·5	6·5
2-Undecanone	8	10	10	30
2-Tridecanone	4	2·6	2·2	8·5
δ-Decalactone		15	19	25
2-Pentadecanone	0·6	0·8	0·8	2·3
δ-Dodecalactone	0·9	4·2	5·5	7

in which the Teflon insert is omitted; sample 4 is a normal indirect UHT milk. It can be seen that for most compounds there is a considerable improvement, which may explain that a laboratory panel significantly differentiated between samples, even if some claimed that the off-flavoured sample was pasteurised milk. This is not such a surprise since it has already been shown that some consumers are rather pleased with a slight cooked flavour.[41]

Lactones are commonly found in cheese. Some correlations have been found in Cheddar and Blue cheese between flavour and the

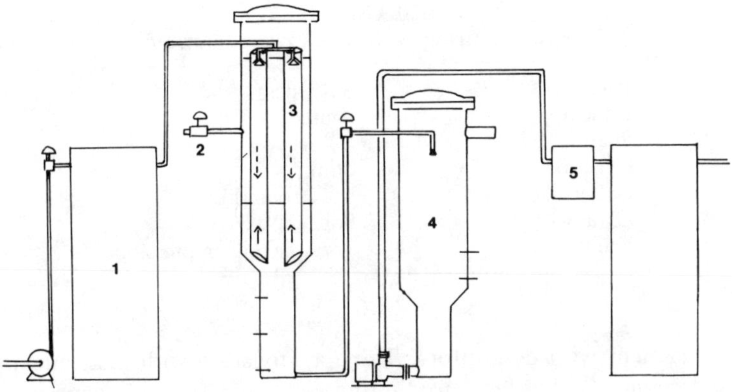

FIG. 2. Modified Laguilharre equipment:[40] 1, milk inlet; 2, steam inlet; 3, Teflon insert; 4, vacuum chamber; 5, homogeniser.

level of lactones, which however seems to increase in rancid Cheddar cheese.[42]

Compounds Arising from Fatty Acids

Methyl ketones (MeK) are common constituents in most dairy products. They may be formed either by decarboxylation of oxo-acids, which are minor constituents of milk fat, or by enzymic oxidative decarboxylation of alkanoic acids. The first mechanism, which is greatly favoured by heating, may explain the increasing level of MeK in milk as the heat treatment becomes more severe (Table 3) or the presence of MeK in some types of cheese.

The second mechanism is of importance for the formation of flavour in mould-ripened cheeses by *Penicillium* species, and above all in Blue cheese where MeK appears as a character impact compound, other important constituents being FFA and secondary alcohols, the latter arising by reduction of the corresponding MeK. The mechanism by which *P. roqueforti* produces MeK has been well described,[43] and the influence of various technological parameters on their appearance in cheese has also been studied.[44] The importance of MeK for flavour imitation explains the attempts to produce Blue cheese flavour or Blue cheese food material by submerged fermentation using lipolysed milk fat or vegetable fat as a substrate.[45–50]

However, MeK does not allow us to explain the variation of total flavour quality between samples. The systematic use of sensory

TABLE 4
Sensory Attributes of Blue Cheese Flavour[51]

Roquefort	Sharpness
Rancid	Ammonia
Soapy	Bitter
Curd-like	Stinky
Butter-like	Mushroom-like
Caramel-like	Sour milk
Fruity	Total flavour impression

assessment with descriptors (Table 4), together with instrumental determination, has led[51] to discriminatory functions allowing us to classify samples according to their quality (Table 5).

Furthermore, multiple regression analysis, via instrumental data, would permit the calculation of total flavour quality (Table 6). This, to my knowledge, is the first attempt to use a statistical approach to explain cheese quality.

Unsaturated ketones have also been found in cheeses, 8-nonen-2-one in Blue cheese[52] and mould surface-ripened cheeses,[53] together with undecenone and tridecenone in Camembert made from milk highly contaminated with psychrophilic bacteria.[54] A mechanism according to which 8-nonen-2-one would result from oxidative decarboxylation of 9-decenoic acid, arising from enzymic oxidative cleavage of polyunsaturated fatty acids, has recently been postulated.[55]

TABLE 5
Calculation of Total Flavour Quality of Blue Cheese[51]

From sensory attributes

$A_1 = 2{\cdot}28$ Roquefort $- 2{\cdot}48$ ammonia $- 1{\cdot}14$ bitter

From instrumental data

$A_2 = 0{\cdot}003\,6$ acetaldehyde $- 0{\cdot}404$ volatile carbonyls $- 0{\cdot}021\,7$ headspace peak 10

	A_1	A_2
First quality	$>3{\cdot}9$	$<-2{\cdot}42$
Second quality	3·9 to 2·0	−2·42 to −1·13
Third quality	$<2{\cdot}0$	$>-1{\cdot}13$

TABLE 6
Calculation of Total Flavour Quality of Blue Cheese by Multiple Regression Analysis with Instrumental Data[51]

$$y = 2{\cdot}505 - 0{\cdot}020 \text{ acidity} - 0{\cdot}246 \text{ isovaleric acid} + 0{\cdot}99 \text{ pH value} - 18{\cdot}07 \text{ SH/FFA} + 1{\cdot}855 \text{ ketones} - 0{\cdot}002 \text{ acetaldehyde} + 0{\cdot}006 \text{ sulphide potential}$$

Compounds Arising from Oxidation

Another mechanism which may lead to flavour compounds is oxidation. However, in most dairy products oxidation leads to off-flavour; it is only in a limited number of cases that oxidation accounts for the formation of volatile compounds which add to the quality.

Oxidation of products such as butter or milk powder fat during storage has been shown to be autocatalytic and the mechanism of the reaction has been clearly demonstrated.[56–58] The influence of different parameters, including added copper content, has been well studied.

The flavour change in whole milk powder during storage, with and without addition of various antioxidants, has recently been re-evaluated.[59] The kinetics of the formation of volatile fat oxidation products have been studied and predictive relationships between flavour characteristics and volatile compounds have been determined.[60,61]

It appears that straight chain aldehydes provide a great contribution to flavour, a result which is in agreement with previous knowledge.

The oxidative deterioration of fat can be monitored by measurement of P.V. or the use of the TBA test.

The measurement of propionaldehyde formed during storage has recently been proposed as an alternative since its level appears to equate directly with the organoleptic evaluation of the development of oxidised flavour, at least in cultured butter.[62]

Oxidation can also be initiated by exposure to natural or artificial light. This may happen in milk and butter. However, the wrapping materials used nowadays for packaging of butter eliminate any risk, while the trend to use polyethylene bottles for milk gives a new actuality to milk photo-oxidation.[63]

In milk the light-induced flavour is related to methional, while in butter the products responsible for off-flavour are those that have been described as contributing to typical oxidation defects,[2] the main one being 1-octen-3-ol, responsible for a mushroom off-flavour.[64]

This compound is found in cheese, where it does not appear as a defect and even brings a large contribution to the characteristic flavour of mould-ripened cheeses such as Camembert[65] and Brie.[66] Too high a level, however, leads to a typical mushroom flavour defect. The origin of 1-octen-3-ol is to be found in the oxidation of linoleic and linolenic acids by *Penicillium caseicolum*. The same mechanism explains the formation of 1,5-octadien-3-one and 1,5-octadien-3-ol, which have also been identified in surface-ripened cheese.[66]

THE ROLE OF PROTEIN IN THE FLAVOUR

Proteins seem to play a capital role in the development of flavour in cheese since proteolysis leads to the formation of important non-volatile and volatile compounds, but probably also because they interact, through active sulphydryl groups, which have been shown to relate closely to flavour formation,[11] with oxido-reduction reactions which could occur during enzymic ripening of milk[11] or as a consequence of lactic acid production.

Peptides

For quite a long period proteins have been considered as a source of small molecules which were contributing to the flavour but mainly to the aroma of dairy products, but the role of proteins and peptides has been ignored by the flavour chemists. The only exception has been the work done on bitter peptides. In cheese, bitterness results from the presence of low molecular weight hydrophobic peptides arising mainly from casein. Among the different caseins α_{s1}-casein produces more bitterness than β-casein. This may explain why goat and ewe milk cheeses are usually devoid of any bitterness as there is relatively little α_{s1}-casein in goat and ewe milk. Several factors[67,68] such as manufacturing conditions, rennet but, above all, cheese microflora have been shown to influence the formation of bitter peptides. It has been suggested that one reason for bitter peptide presence in cheese could be the inability of peptidase-deficient starter strains to degrade the bitter peptides to non-bitter peptides and amino acids. Another reason, which of course only applies to mould-ripened cheese, is that *P. caseicolum* is itself able to cause bitterness if its growth is significant. This happens when large amounts of lactic acid are produced by a highly active starter strain, which favours the activity of the mould. This is demonstrated by reducing the occurrence of bitterness by slowing down the growth of the *Penicillium*, which can be done by raising the pH of the cheese, either by using an ammoniacal atmosphere[69] or by inoculating *Geotrichum candidum*.[70]

Bitterness has gained a new importance since various bacterial or fungal proteinases[31] have experimentally been used to accelerate ripening in cheeses. Most efforts have been impeded by the difficulty of obtaining a balanced flavour which, if present, is often impaired by an excessive bitterness.[71–73]

The importance of peptides was recognised when the non-volatile water-soluble fraction was studied in Swiss cheese.[74] Results suggest that the typical sweet flavour arises from an interaction of calcium and magnesium with small peptides, while the brothy-nutty flavour is found to be related to small peptides and amino acids, and the burned-bitter flavour to medium sized (tri- to hexa-) peptides. This burned note, which we ourselves observed on Gruyère protein remaining after supercritical CO_2 extraction of fat, has also been isolated as a fraction from *Lactobacillus helveticus* culture and attributed to a complex between amino acids and 2,5-dimethyl-4-hydroxy-3-[2H]furanone.[77] Similar experiments on Cheddar[78] have shown that the non-volatile water-extractable components make a significant direct contribution by increasing flavour intensity. Other observations suggest that an important factor is the effect of proteolysis on the binding of volatile compounds because binding decreases as proteolysis increases,[75] but we still know very little of casein degradation and/or the respective roles of the different caseins in the ripening process. Progress could be made on that point in the near future as it is now possible to prepare β-casein on a large scale using microfiltration.[79] This could allow the influence of each casein to be observed more easily than with the use of milk from genetically different animals.[80]

Amino Acids

It is difficult to assess the importance of free amino acids in cheese despite, or because of, the numerous available results, mainly obtained on Cheddar,[81] which are not always consistent. Amino acids, however, play a key role as precursors of a large number of flavour compounds.

Amino Acids as a Source of Flavour Compounds

Ammonia, amines and amides

Very little attention has been given to ammonia which could be of importance in some types of cheese, not only by the rise of pH it causes, but also by its own flavour character. The same applies to amines which have been identified in several types of cheese[82–84] but for which no quantitative data are given. Amides have also been repeatedly identified,[84,85] but again we lack information on their possible implication in flavour.

Aldehydes, alcohols and phenols

Aldehydes may appear in cheese together with alcohols from amino acids via the Strecker degradation. In Cheddar some of these compounds have been held responsible for 'unclean' flavours.[86] Phenylethanol and phenylacetaldehyde give unpleasant flavours, the latter also contributing to an astringent bitterness; 3-methylbutanal, 2-methylbutanal, 2-methylpropanal and methional have been described as giving unclean, harsh and dull flavour, while cresol is held responsible for a typical barley note. This compound has already been shown to be responsible for flavour defects in Gouda[87] and surface-ripened cheeses,[88] but it has also been identified in good quality cheeses, being then presumably present at lower level or having its effect balanced by other components.[89]

Phenol, which has been considered as contributing to the quality of Cheddar,[90] appears as the major flavour compound in surface-ripened cheeses.[90] It is interesting to note that small concentrations of phenol and *p*-cresol have been shown to be part of the desirable flavour of sweet cream butter.[91] The same applies to indole[91] which has also been identified in surface-ripened cheeses[90] and Gruyère.[92]

Branched chain fatty acids

Strecker aldehydes may be further oxidised to acids. This mechanism explains the occurrence of volatile iso-acids which have been repeatedly identified in surface-ripened cheeses (Fig. 3).[84,90,93] These acids, which have a lower threshold value than the straight chain ones, could be of importance in the flavour of these cheeses and even be key compounds as in Tilsiter.[84]

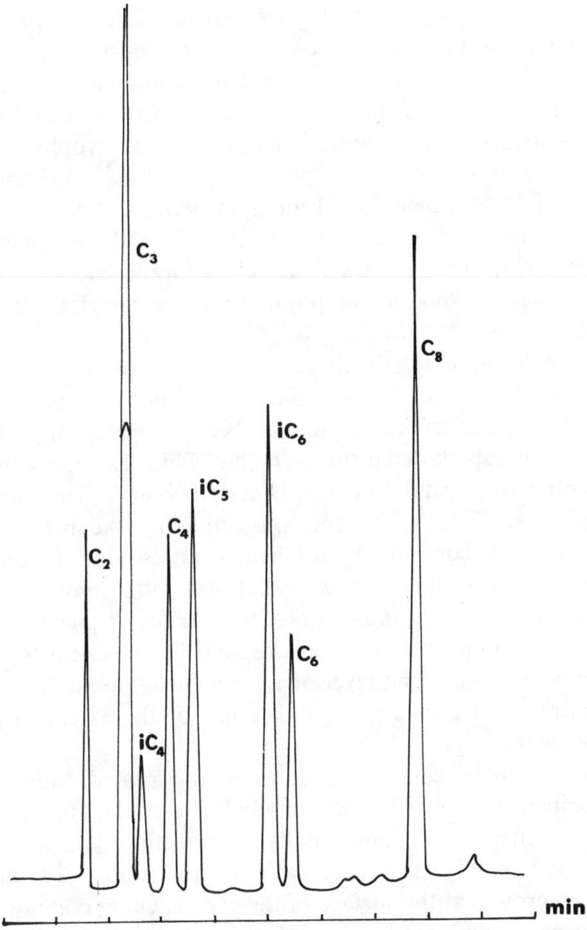

FIG. 3. An example of free fatty acids in surface-ripened cheeses (Pont l'Eveque).[93] Teflon column 3 ft × $\frac{1}{8}$ in; Chromosorb G + 5 p. 100 C20M + 0·25 p. 100 terephthalic acid; temp. programmed 60–160°C at 2°/min.

Sulphur compounds

Of special interest are the decomposition products of the sulphur amino acids cysteine and methionine. These products are in some cases responsible for typical flavour defects while in most cases they contribute to the typical flavour of products.

Hydrogen sulphide, which is formed in milk upon heating, is responsible for the cooked flavour that is observed in UHT milk,[2] and sometimes in butter.[2] Upon severe treatment, other sulphur components such as methanethiol may also be formed.

Dimethyl sulphide which is a normal component of raw milk may, if present at too high a level, lead to an unclean flavour as happens in milk highly contaminated with psychrotrophic bacteria.[2]

In cheese itself sulphur compounds may also be responsible for taints or defects; 4-mercapto-4-methylpentanone-2 has been shown to be responsible for catty taint in Gouda,[94] and methanethiol has been shown to give a marked defect to Grana cheese.[95] In most cases, however, sulphur compounds greatly contribute to cheese flavour.

Hydrogen sulphide, which is present in any type of matured cheese, has been described by some as an important compound in Cheddar, while others conclude that its role is not so important. Its role is largely superseded by methanethiol which has been considered as essential to normal Cheddar cheese flavour.[96] This opinion is not, however, held by others who, studying the production of volatile sulphur compounds in Cheddar cheese during accelerated ripening, came to the conclusion that none of the volatile sulphur compounds are reliable indicators of flavour development.[97] Despite correlations, both methanethiol level and flavour seem to increase independently, while the level of PTA-soluble amino N seems, on the contrary, to give a good indication of the extent of flavour development.[98]

Several sulphur compounds have been repeatedly identified in mould-ripened cheese:[99] H_2S, dimethyl disulphide, dimethyl trisulphide and methionol. Additionally methanethiol is also found in Camembert, mainly when there is a *Bacterium linens* or a *Geotrichum candidum* growth at the surface of the cheese. Other compounds are also normally present, some of them as traces: 2,4-dithiapentane (which has also been identified in Gouda[100]); 3,4-dithiahexane; 2,4,5-trithiahexane; 3-methylthio-2,4-dithiapentane. The exact role of each of these compounds remains to be clarified.

From the results recently obtained on Limburger[90] and those obtained earlier on some other surface-ripened cheeses,[85] a specific role can be attributed to methanethiol, which is partly converted to dimethyl disulphide and higher polysulphides by oxidation or esterified to methylthioesters by volatile FFA.

Methanethiol and its esters are also present in Gruyère together with dimethyl disulphide and dimethyl sulphide. The latter, the concentration of which increases as one goes deeper into the cheese, is known to be a metabolite of propionibacteria together with 2-methyl-tetrahydrothiophen-3-one,[84] a compound also identified as a trace in acid cream butter.[17]

Pyrazines

There is now strong evidence for the presence of pyrazines in dairy products but, except in a few cases where they have been shown to be responsible for a typical flavour defect, their real importance has not been clearly demonstrated, even if their very low threshold values suggest that they could be important.

Dumont *et al.*[101] identified dimethylpyrazine in grated Parmesan cheese, then several alkylpyrazines were identified in Emmental cheese[102] and in Gouda.[100] In the same period, acetylpyrazine and 2-methoxy-3-ethylpyrazine have been considered as of some importance for the flavour of old Cheddar cheese.[89] Since then, pyrazines have been identified in processed American cheese where they appear as likely contributors of the weak nutty note,[103] surface-ripened cheeses,[104] Gruyère[105,106] and recently in cultured cream butter.[17]

The possible microbial origin of pyrazines became evident since Morgan[107] isolated (from milk contaminated by *Pseudomonas taetrolens*) dimethylpyrazine and 2-methoxy-isopropylpyrazine (MIP), the latter being responsible for a strong potato-like defect. The same bacteria were later isolated from a surface-ripened cheese presenting the same defect for which MIP was also found responsible.[108] The microbial origin of pyrazines is again suggested by the observation of Liardon *et al.*[92] who noticed that alkylpyrazines were only to be found in the outer zone of Gruyère cheese where intense microbiological activity takes place during maturation. Their level seems to increase linearly during ripening.[109] *Penicillium caseicolum* cultures have been shown to be able to produce MIP, which at low concentration could contribute to the flavour of mould-ripened cheeses,[110] while *Lactobacillus helveticus* cultures were shown to produce alkylpyrazines.[77]

The pathways which lead to pyrazine formation in dairy products are not clear as *P. taetrolens* is able to produce MIP with nitrate as sole source of nitrogen.[107] The results obtained by Gallois,[111] who studied MIP biosynthesis using labelled amino acids, suggest that pyrazine could be formed by amino acid condensation.

CONCLUSIONS

The knowledge accumulated in various related fields such as dairy microbiology, technology and flavour chemistry has permitted us to gain some understanding of the more salient phenomena, but the ultimate understanding is missing. Some progress can yet be made by flavour chemists with the hope to differentiate analytically between samples that are recognised to be of the same quality but are perceived as different on tasting. This does not probably imply much effort to identify unknown volatile compounds but rather to obtain quantitative data. At the same time the influence of non-volatile compounds should be more thoroughly studied, which implies analytical chemistry but also sensory analysis. We have seen that some efforts have been made towards a better definition of what we call flavour through use of descriptive profiling, which will need some efforts on vocabulary. We all know we can be very precise in describing defects and how short of words we are when we have to describe a perfect sample. One can also wonder what would be the benefit of modern statistical techniques to try to extract relevant information from the mass of obtainable results. No one would be able to do so alone because of the number of analyses required. This probably explains why so many of us have too often tried to use an individual flavour component as an index of quality with the risks that this implies. The statistical approach would need a cooperative action between institutes on a common model cheese. Finally, as suggested by Kristoffersen,[11] we should probably attach more thought to milk itself.

REFERENCES

1. Behnke, U. (1980). *Nährung*, **24**, 71–83.
2. Badings, H. T. and Neeter, R. (1980). *Neth. Milk Dairy J.*, **34**, 9–30.
3. Law, B. A. (1981). *Dairy Sci. Abstr.*, **43**, 143–54.
4. Adda, J., Gripon, J. C. and Vassal, L. (1982). *Food Chem.*, **9**, 115–29.
5. Manning, D. J. and Nursten, H. E. (1985). In: *Developments in Dairy Chemistry*, Vol. 3, Ed. Fox, P. S., Elsevier Applied Science, London, 217–38.
6. Lawrence, R. C., Gilles, J. and Creamer, L. K. (1983). *New Zealand J. Dairy Sci. Technol.*, **18**, 175–90.

7. Lawrence, R. C., Heap, H. A. and Gilles, J. (1984). *J. Dairy Sci.*, **67**, 1632–45.
8. Mocquot, G. (1971). *Proc. Int. Symp. on Conversion and Manufacture of Foodstuffs by Microorganisms*, Kyoto, Japan, 191–7.
9. Law, B. A., Castanon, M. J. and Sharp, M. E. (1976). *J. Dairy Res.*, **43**, 301–11.
10. Mannings, D. J. (1978). *Dairy Ind. Int.*, **43**, 37–9.
11. Kristoffersen, T. (1985). *Milchwissenschaft*, **40**, 197–9.
12. Babel, F. J. and Hammer, B. W. (1939). *Food Res.*, 81–5.
13. Steinsholt, V. K., Svensen, A. and Tufto, G. (1971). Report No. 155, Dairy Research Institute, Agricultural College of Norway.
14. Robinson, R. K., Tamine, A. Y. and Chubb, L. W. (1977). *Milk Ind.*, **79**, 4–6.
15. Green, M. L. and Manning, D. J. (1982). *J. Dairy Res.*, **49**, 737–48.
16. Collins, E. B. (1972). *J. Dairy Sci.*, **55**, 1022–8.
17. Mick, S., Mick, W. and Schreier, P. (1982). *Milchwissenschaft*, **37**, 661–5.
18. Law, B. A. (1981). *Dairy Sci. Abstr.*, **43**, 143–54.
19. Langler, J. E., Libbey, L. M. and Day, E. A. (1967). *J. Agric. Food Chem.*, **15**, 386–91.
20. Biede, S. L. and Hammond, E. G. (1979). *J. Dairy Sci.*, **62**, 238–48.
21. Magee, E. L. Jr, Olson, N. F. and Lindsay, R. C. (1981). *J. Dairy Sci.*, **64**, 616–21.
22. Keen, A. R., Walker, N. J. and Pederby, M. F. (1974). *J. Dairy Res.*, **41**, 249–54.
23. Foda, E. A., Hammond, E. G., Reinbold, G. W. and Hotchkiss, D. K. (1974). *J. Dairy Sci.*, **57**, 1137–42.
24. Bakkene, G. and Steinsholt, K. (1975). Report No. 189, Dairy Research Institute, Agricultural College of Norway.
25. Bakke, H., Steine, T. and Eggum, A. (1977). *Acta Agric. Scand.*, **27**, 245–9.
26. Smith, P. W., Parks, O. W. and Schwartz, D. P. (1984). *J. Dairy Sci.*, **67**, 794–801.
27. Ramshaw, E. H. (1974). *Austr. J. Dairy Technol.*, **29**, 110–14.
28. Woo, A. H. and Lindsay, R. C. (1983). *J. Food Sci.*, **48**, 1761–6.
29. Law, B. A. and Wigmore, A. S. (1985). *J. Soc. Dairy Technol.*, **38**, 86–7.
30. Kilara, A. (1985). *Process Biochem.*, **20**, 35–44.
31. Sood, V. K. and Kosikowski, F. V. (1979). *J. Dairy Sci.*, **62**, 1865–72.
32. Huang, H. T. and Dooley, J. G. (1976). *Biotechnol. Bioeng.*, **18**, 909–19.
33. Dimick, P. S., Walker, N. J. and Patton, S. (1969). *J. Agric. Food Chem.*, **17**, 649–55.
34. Urbach, G. (1982). *Lebensm. Wiss. Technol.*, **15**, 62–7.
35. Kinsella, J. E., Patton, S. and Dimick, P. S. (1967). *J. Am. Oil Chem. Soc.*, **44**, 449–54.
36. Ellis, R. and Wong, N. P. (1975). *J. Am. Oil Chem. Soc.*, **52**, 252–5.

37. STARK, W. and URBACH, G. (1976). *Austr. J. Dairy Technol.*, **31**, 80–2.
38. JEON, I. J. (1976). *Identification of Volatile Flavor Compounds and Variables affecting the Development of Off-flavor in Ultra-high-temperature Processed Sterile Milk*, University of Minnesota.
39. BADINGS, H. T., VAN DER POL, J. J. G. and NEETER, R. (1981). In: *Flavour '81*, Ed. Schreier, P., de Gruyter, Berlin, 683–92.
40. LAITERIE ST DENIS L'HOTEL. (1985). Fr. Pat. 85,106,58.
41. DEANE, D. D., CHELESVIG, J. A. and THOMAS, W. R. (1967). *J. Dairy Sci.*, **50**, 1216–20.
42. WONG, N. P., ELLIS, R. and LA CROIX, D. E. (1975). *J. Dairy Sci.*, **58**, 1437–41.
43. HAWKE, J. C. (1966). *J. Dairy Res.*, **33**, 225–41.
44. GODINHO, M. and FOX, P. F. (1981). *Milchwissenschaft*, **36**, 476–8.
45. DWIVEDI, B. K. and KINSELLA, J. E. (1974). *J. Food Sci.*, **39**, 620–3.
46. JOLLY, R. and KOSIKOWSKI, F. V. (1975). *J. Food Sci.*, **40**, 285–7.
47. JOLLY, R. C. and KOSIKOWSKI, F. V. (1975). *J. Dairy Sci.*, **58**, 1272–5.
48. KNIGHT, S. G. (1963). US Pat. 3,100,153.
49. KANISAWA, T. and ITOH, H. (1984). *Nippon Shokuhin Kogyo Gakkaishi*, **31**, 483–7.
50. NELSON, J. H. (1970). *J. Agric. Food Chem.*, **18**, 567–9.
51. ROTHE, M., ENGST, W. and ERHARDT, V. (1982). *Nährung*, **26**, 591–602.
52. SVENSEN, A. and OTTESTAD, E. (1969). Report No. 141, Dairy Research Institute, Agricultural College of Norway.
53. DUMONT, J. P., ROGER, S., CERF, P. and ADDA, J. (1974). *Lait*, **54**, 501–16.
54. DUMONT, J. P., DELESPAUL, G., MIGUOT, B. and ADDA, J. (1977). *Lait*, **57**, 619–30.
55. KARAHADIAN, C., JOSEPHSON, D. B. and LINDSAY, R. C. (1985). *J. Dairy Sci.*, **68**, 1865–77.
56. STARK, W. and FORSS, D. A. (1964). *J. Dairy Res.*, **31**, 253–9.
57. STARK, W. and FORSS, D. A. (1966). *J. Dairy Res.*, **33**, 31–6.
58. BADINGS, H. T. (1970). *Cold Storage Defects in Butter and their Relation to the Autoxidation of Unsaturated Fatty Acids*, University of Wageningen.
59. HALL, G. and LINGNERT, H. (1983). *J. Food Quality*, **7**, 131–51.
60. HALL, G., ANDERSSON, J., LINGNERT, H. and OLOFSSON, B. (1985). *J. Food Quality*, **9**, 153–90.
61. HALL, G. and ANDERSSON, J. (1985). *J. Food Quality*, **7**, 237–53.
62. TANIMURA, S., NAKABAYASHI, A. and KATAYAMA, H. (1982). *New Zealand J. Dairy Sci. Technol.*, **17**, 47–53.
63. SCHRÖDER, M. J. A., SCOTT, K. J., BLAND, M. A. and BISHOP, D. R. (1985). *J. Soc. Dairy Technol.*, **38**, 48–52.
64. FOLEY, J., O'DONOVAN, D. and COONEY, C. (1971). *J. Soc. Dairy Technol.*, **24**, 38–44.
65. MOINAS, M., GROUX, M. and HORMAN, I. (1973). *Lait*, **53**, 601–9.
66. KARAHADIAN, C., JOSEPHSON, D. B. and LINDSAY, R. C. (1985). *J. Dairy Sci.*, **68**, 1865–7.

67. Czulak, J. (1959). *Austr. J. Dairy Technol.*, **14**, 177–9.
68. Lawrence, R. C., Creamer, L. K., Gilles, J. and Martley, F. G. (1972). *New Zealand J. Dairy Sci. Technol.*, **7**, 32–7.
69. Vassal, L. and Grippon, J. C. (1984). *Lait*, **64**, 397–417.
70. Mourgues, R., Bergere, J. L. and Vassal, L. (1983). *Tech. Lait.*, **978**, 11–15.
71. Law, B. A. and Wigmore, A. (1982). *J. Dairy Res.*, **49**, 137–46.
72. Law, B. A. (1980). *Dairy Ind. Int.*, **45**, 17–22, 48.
73. Law, B. A. (1983). *IDF Bull.*, **157**, 33–9.
74. Biede, S. L. (1977). *A Study of the Chemical and Flavor Profiles of Swiss Cheese*, Iowa State University.
75. Biede, S. L. and Hammond, E. G. (1979). *J. Dairy Sci.*, **62**, 227–37.
76. Biede, S. L. and Hammond, E. G. (1979). *J. Dairy Sci.*, **62**, 238–48.
77. Kowalewska, J., Zelazowska, H., Babuchowski, A., Hammond, E. G., Glatz, B. A. and Ross, F. (1985). *J. Dairy Sci.*, **68**, 2165–71.
78. McGugan, W. A., Emmons, D. B. and Larmond, E. (1979). *J. Dairy Sci.*, **62**, 398–403.
79. Terre, E., Maubois, J. L., Brule, G. and Pierre, A. (1986). Fr. Pat. 86,00,325.
80. Vassal, L., to be published.
81. Aston, J. W. and Dulley, J. R. (1982). *Austr. J. Dairy Technol.*, **37**, 59–64.
82. Ney, K. H. and Wirotama, I. P. G. (1971). *Z. Lebensm. Unters. Forsch.*, **146**, 343–4.
83. Ney, K. H. and Wirotama, I. P. G. (1978). *Fette Seifen Anstr.*, **80**, 249–51.
84. Ney, K. H. and Wirotama, I. P. G. (1985). *Fette Seifen Anstr.*, **87**, 289–94.
85. Adda, J., Roger, S. and Dumont, J. P. (1978). In: *Flavor of Foods and Beverages: Chemistry and Technology*, Eds Charalambous, G. and Inglett, G. E., Academic Press, New York, 67.
86. Dunn, H. C. and Lindsay, R. C. (1985). *J. Dairy Sci.*, **68**, 2859–74.
87. Badings, H. T., Stadhouders, J. and Van Duin, H. (1968). *J. Dairy Sci.*, **51**, 31–5.
88. Dumont, J. P., Roger, S., Cerf, P. and Adda, J. (1974). *Lait*, **54**, 243–51.
89. McGugan, W. A. (1975). *J. Agric. Food Chem.*, **23**, 1047–50.
90. Parliment, T. H., Kolor, M. and Rizzo, D. J. (1982). *J. Agric. Food Chem.*, **30**, 1006–8.
91. Urbach, G., Stark, W. and Forss, D. A. (1972). *J. Dairy Res.*, **39**, 35–47.
92. Liardon, R., Bosset, J. O. and Blanc, B. (1982). *Lebensm. Wiss. Technol.*, **15**, 143–7.
93. Stark, W. and Adda, J. (1972). *Tech. Lait.*, **746**, 15–17.
94. Badings, H. T. (1967). *J. Dairy Sci.*, **50**, 1347–51.
95. Battistotti, B., Bosi, F., Scolari, G. L., Bottazzi, V., Chiusa, P., Gonzaga, E. and Pecorari, M. (1985). *Sci. Tec. latt.-casear*, **36**, 77–97.

96. Manning, D. J. (1979). *J. Dairy Res.*, **46**, 531–7.
97. Aston, J. W. and Douglas, K. (1983). *Austr. J. Dairy Technol.*, **38**, 66–70.
98. Aston, J. W., Durward, I. G. and Dulley, J. R. (1983). *Austr. J. Dairy Technol.*, **38**, 55–9.
99. Dumont, J. P., Roger, S. and Adda, J. (1976). *Lait*, **56**, 1–5.
100. Slott, D. and Harkes, P. D. (1975). *J. Agric. Food Chem.*, **23**, 356–7.
101. Dumont, J. P., Roger, S. and Adda, J. (1974). *Lait*, **537**, 386–96.
102. Slott, D. and Hofman, H. J. (1975). *J. Agric. Food Chem.*, **23**, 358.
103. Lin, S. S. (1976). *J. Agric. Food Chem.*, **24**, 1252–4.
104. Dumont, J. P., Roger, S., Cerf, P. and Adda, J. (1974). *Lait*, **54**, 243–51.
105. Dumont, J. P. and Adda, J. (1978). *J. Agric. Food Chem.*, **26**, 364–7.
106. Dumont, J. P., Adda, J. and Rousseaux, P. (1981). *Lebensm. Wiss. Technol.*, **14**, 198–202.
107. Morgan, M. E. (1976). *Biotechnol. Bioeng.*, **18**, 953–65.
108. Dumont, J. P., Mourgues, R. and Adda, J. (1983). In: *Sensory Quality of Food and Beverages: Definition, Measurement and Control*, Eds Williams, A. A. and Atkin, R. K., Ellis Horwood, Chichester, 424–8.
109. Bosset, J. O. and Liardon, R. (1985). *Lebensm. Wiss. Technol.*, **18**, 178–85.
110. Karahadian, C., Josephson, D. B. and Lindsay, R. C. (1985). *J. Agric. Food Chem.*, **33**, 339–43.
111. Gallois, A. (1984). *Biosynthèse de la Methoxy-2-isopropyl-3-pyrazine*, Institut National Agronomique, Paris.

11

Aroma Compounds from the Maillard Reaction

H. E. NURSTEN

Department of Food Science, University of Reading, UK

ABSTRACT

The complexities of the Maillard reaction are outlined, as background to considering the types of aroma compound produced. Recent work is then reviewed. Much of it is concerned with model systems. The amino acids involved in these form a framework for the discussion: proline, hydroxyproline, serine and threonine, cystine and cysteine, phenylalanine, leucine, and glycine. Stress is laid on the need to understand the balance between 1,2- and 2,3-enolisation of Amadori compounds and between aldolisation and dealdolisation, the effects of oligomers in contrast to monosaccharides and amino acids, and the interactions with other food components, such as lipids.

INTRODUCTION

The Maillard reaction is still best summarised by the scheme due to Hodge,[1] illustrated in Fig. 1. It is worth noting:

(1) Sugars on their own, provided they are exposed to higher temperatures, will undergo similar reactions by themselves (caramelisation), but then, in the extreme, there can be no interaction with amino compounds and no nitrogen-containing compounds can result.

(2) The Maillard reaction (and caramelisation) is best divided into three stages. The initial one comprises glycosylamine formation (A) and subsequent rearrangement (B). The intermediate stage

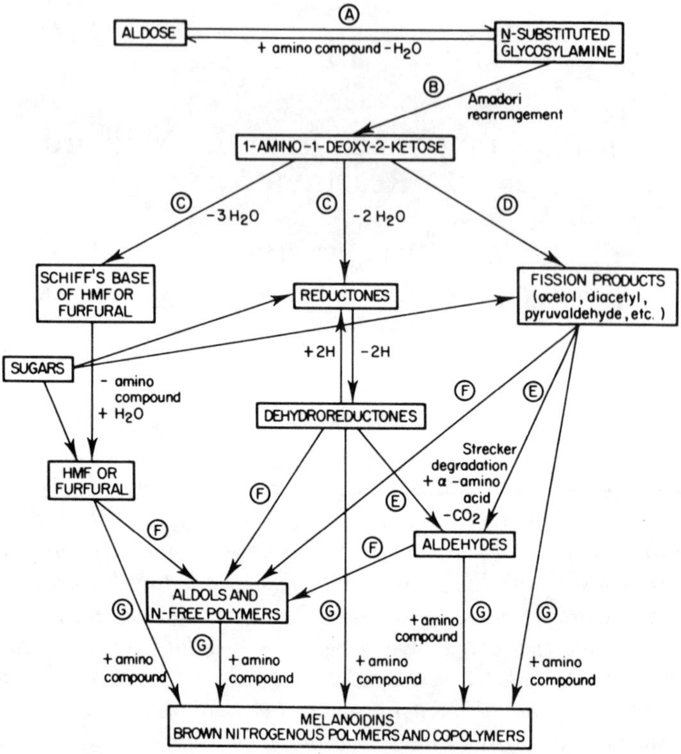

FIG. 1. Non-enzymic browning.

comprises dehydration (C), either by loss of three molecules of water to furfurals or by loss of two to reductones; fission (D), mainly by dealdolisation; and Strecker degradation (E), the interaction of amino acids and dicarbonyl compounds, which may be either dehydroreductones or dehydration or fission products. The final stage consists of the conversion of carbonyl compounds, be they furfurals, fission products, dehydroreductones or Strecker aldehydes, into high molecular weight products, the melanoidins, with further involvement of amines, where these are available.

(3) Ketoses undergo similar reactions, reaction B then being termed the Heyns rearrangement.

(4) Some reactions are reversible, notably the first step A and the dehydrogenation of reductones.

The Maillard reaction is central to food chemistry because of at least five aspects:

(1) Production of colour. This may be desirable, as in coffee or bread crust, or undesirable, as in glucose syrup and in many intermediate moisture products. Colour is principally produced in the final stage.

(2) Production of flavours or off-flavours. These form the theme of this paper. Odours and off-odours are due to volatile products, e.g. fission products and Strecker aldehydes. Substances tasting sweet or bitter may also be involved, as well as effects on texture. Volatiles are produced in the intermediate and final stages.

(3) Reduction in nutritional value by involvement of ascorbic acid (a reductone) and of lysine (free or bound), an essential and often limiting amino acid. Metal-chelating properties may also be significant.

(4) Toxicity through possible formation of imidazoles and of *N*-nitroso derivatives, e.g. Amadori compounds. The intrinsic toxicity of Maillard products and intermediates is of some concern, but is difficult to study because of their anti-nutritional properties.

(5) Antioxidant properties. These are thought to be due to the reductones formed, but chelation of heavy metals may also be involved.

From the above, it will be clear that the flavour scientist cannot properly pursue his interest in the Maillard reaction without bearing its other aspects very much in mind.

Here we are mainly concerned with the aroma compounds produced by the Maillard reaction. They are best classified into three groups:[2]

(1) 'Simple' sugar dehydration/fragmentation products:
Furans
Pyrones
Cyclopentenes
Carbonyls
Acids

(2) 'Simple' amino acid degradation products:
Aldehydes
Sulphur compounds

(3) Volatiles produced by further interactions:
- Pyrroles
- Pyridines
- Imidazoles
- Pyrazines
- Oxazoles
- Thiazoles
- Compounds from aldol condensations

Figures 2–4 will serve as a reminder of important volatiles of these three classes.

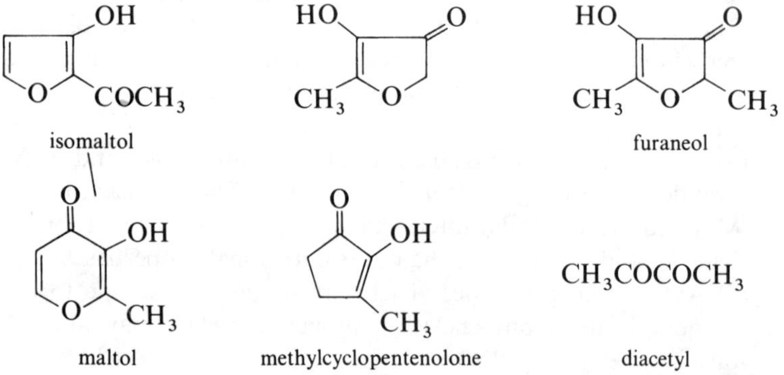

FIG. 2. Important 'simple' sugar dehydration/fragmentation volatiles.

The above was intended as a brief summary of the current view of volatiles produced by the Maillard reaction. Several more extensive reviews are available.[3–6]

The major part of this paper remains to be devoted to recent work in this area. Reporting on this work is made much easier by the fact that the Third International Symposium on the Maillard Reaction[7] was held in Japan in July 1985, following the earlier symposia in Sweden[8] and Las Vegas.[9]

Recent work is best dealt with according to the amino acid studied, much of the research being based on model systems, rather than foods. The amino acids concerned are: proline, hydroxyproline,

$$(CH_3)_2CHCH_2CHO \qquad H_2S$$
$$\text{etc.} \qquad CH_3SH$$

FIG. 3. Important 'simple' amino acid degradation volatiles.

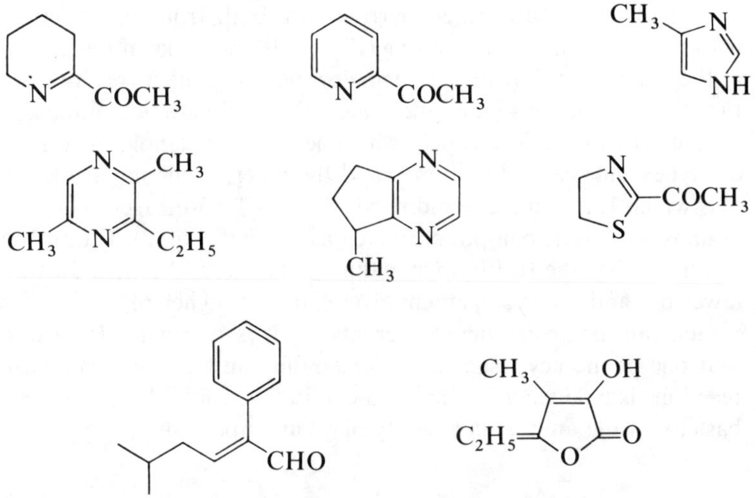

FIG. 4. Significant volatiles produced by further interactions.

serine and threonine, cystine and cysteine, phenylalanine, leucine and glycine. Tressl's group[10] has done outstanding work, which is properly given pride of place.

PROLINE

Proline along with hydroxyproline, being imino acids, clearly differ from amino acids. They cannot part with ammonia in the Strecker degradation and so do not give rise to that important group of volatiles, the pyrazines. Nevertheless proline plays an outstanding role in the Maillard reactions that occur in foods. More than 120 proline-specific compounds have been identified, depending on the reducing sugar and the reaction conditions.

Tressl *et al.*[11] heated proline separately with glyceraldehyde, erythrose, arabinose, glucose and rhamnose at 150°C for 30 min and analysed the volatiles formed. They will be considered here in three groups: pyrrolidines and piperidines, pyrrolizines, and azepines.

Pyrrolidines and Piperidines

The volatiles included twelve *N*-substituted pyrrolidines and four 2-substituted piperidines, as well as l-pyrroline.

Two of the *N*-substituted pyrrolidines, both from the reaction with rhamnose, are worth noting (Fig. 5). If one looks more closely at the reactions following the Amadori rearrangement (reaction B, Fig. 1), as set out in Fig. 6, one can see that there are two branches, one, occurring at lower pH, which leads to melanoidins via 3-deoxyhexosuloses and furfurals, and the other, occurring at higher pH, which leads to melanoidins via 1-methyl-2,3-dicarbonyl compounds and cyclic compounds (such as those in Fig. 2). From this it follows that the furfurylpyrrolidine is the more likely product at lower pH and the cyclopentenylpyrrolidine at higher pH.

Generalising from these observations, it is worth pointing out that one of the key issues in understanding how to steer Maillard reactions is balancing 1,2- and 2,3-enolisation correctly. In this the basicity of the amino acid clearly plays an important role.

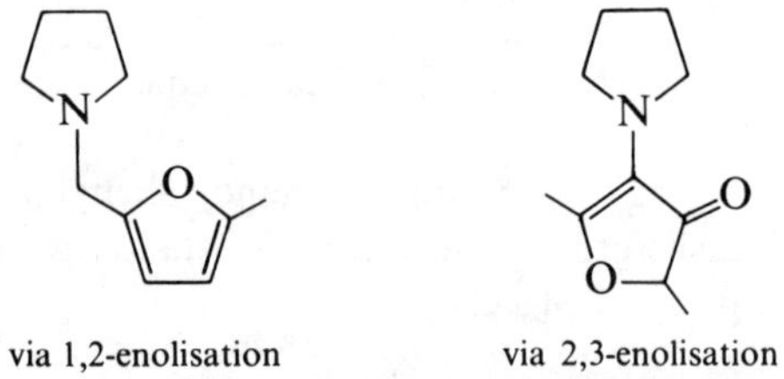

FIG. 5. Two *N*-substituted pyrrolidines from proline–rhamnose interaction.

The pyrrolidines possess cereal aromas and the *N*-furfuryl derivative (Fig. 5) and its lower homologue have sesame-like notes.

The piperidines possess no characteristic odours, but are easily transformed into the corresponding tetrahydropyridines with bready, cracker-like odour qualities and thresholds near 1 in 10^9. Tressl *et al.*[10] consider that ring enlargement of pyrrolidines to piperidines could well take place as in Fig. 7.

It is worth remembering that in foods pyridines can also result from the interaction of amino acids with alkanals, as shown by Suyama and Adachi.[12] An additional pyridine, the 2-isobutyl-3,5-diisopropyl derivative, was recently identified in fried chicken flavour, as well as 3,5-diisobutyl-1,2,4-trithiolane.[13] It had a roasted, cocoa-like aroma. Both compounds, as well as others, also resulted from the interaction of isovaleraldehyde and ammonium sulphide, the trithiolane, as well as 5,6-dihydro-2,4,6-triisobutyl-4*H*-1,3,5-dithiazine, being described as possessing fried bacon aroma.[14]

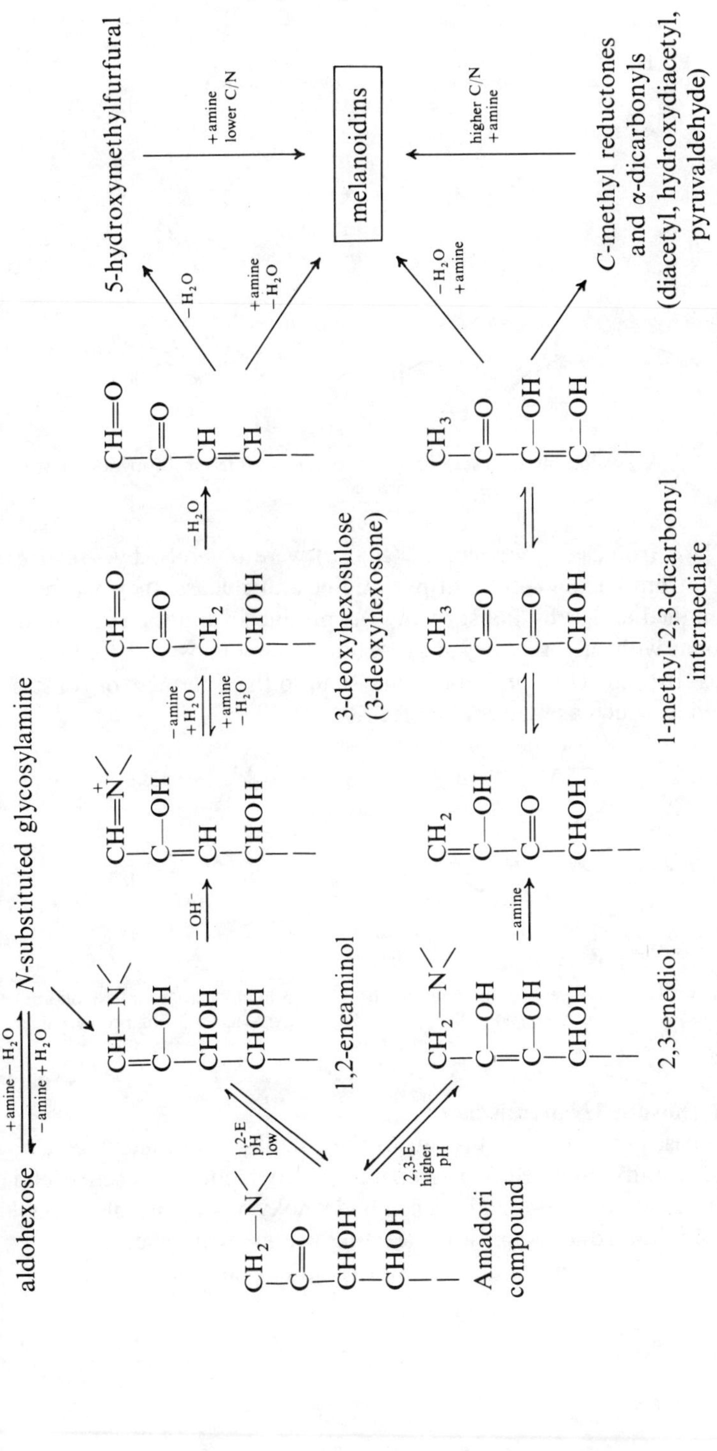

FIG. 6. Maillard reaction: two major pathways from Amadori compounds to melanoidins (based on Hodge[1]).

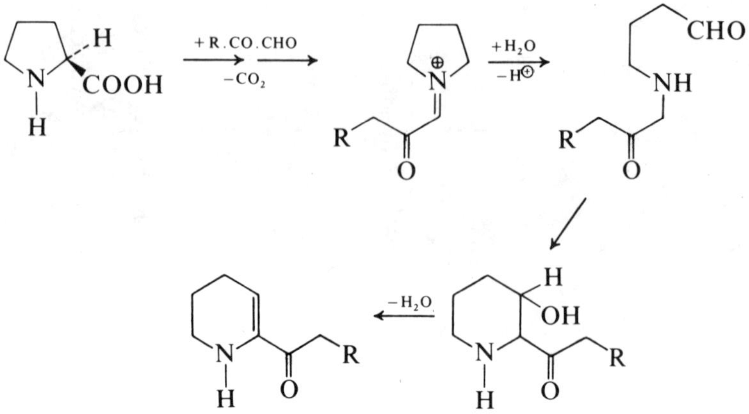

FIG. 7. A possible mechanism for the formation of tetrahydropyridines from proline.

Two trinuclear pyrrolidines (Fig. 8) were obtained by Tressl *et al.*[11] from the interaction of pyrrolidine and glucose; they had been identified earlier by Pabst *et al.*[15] as products from the reaction of proline with sucrose. They are very bitter compounds, with thresholds of 5–10 ppm. They probably contribute to the bitterness of roasted products such as malt and coffee.

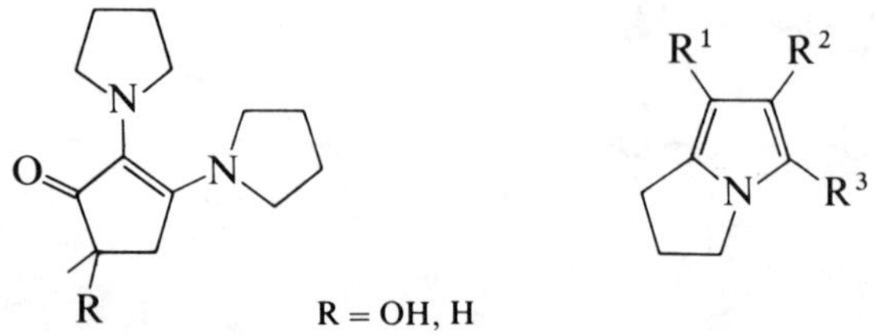

FIG. 8. Left: two very bitter compounds from proline–sucrose or pyrrolidine–glucose interaction (threshold 5–10 ppm). Right: pyrrolizines from proline–sugar interaction.

2,3-Dihydro-1*H*-pyrrolizines

These are major constituents of the volatiles. Shigematsu *et al.*[16] had already obtained 3 pyrrolizines by interaction of proline and glucose at 200°C for 5 min. They had smoky, roasty aromas. Tressl *et al.*[17] added another 18 by reacting proline with glyceraldehyde, erythrose, arabinose, glucose and rhamnose at 150°C for 90 min.

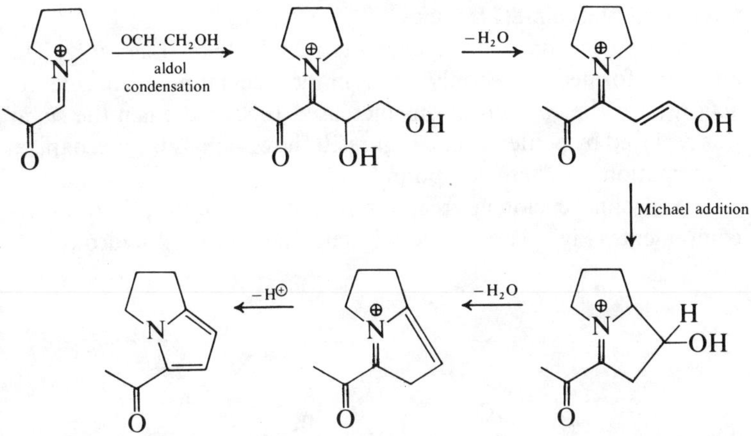

FIG. 9. A possible mechanism for the formation of pyrrolizines from proline.

The general formula for these pyrrolizines is included in Fig. 8. Tressl *et al.*[17] picture their formation as in Fig. 9.

The pyrrolidine nucleus of proline remains hydrogenated and both aldol condensations and dealdolisation must be taking place since, although glyceraldehyde is used, a fragment of greater than C_3 is added, in fact $2 \times C_3$–C_1.

Generalising again, it is worth stressing that another of the key issues in understanding how to steer Maillard reactions is balancing aldol condensation and the retro reaction.

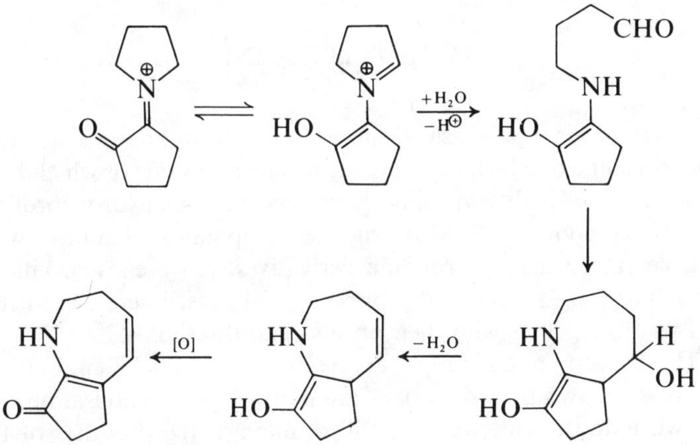

FIG. 10. A possible mechanism for the formation of azepines from proline.

Cyclopent[*b*]azepin-8(1*H*)-ones

Ten azepines have been characterised by Tressl *et al.*[18] The amounts formed were only ppm in proline–monosaccharide reactions, but the concentrations increased 1000-fold when the sugar was replaced by pentenolone. Figure 10 shows a possible mechanism of formation for these compounds.

The proline–cyclotene reaction product is bitter, two of its components (Fig. 11) being very bitter. The proline–2-hydroxy-3,4-

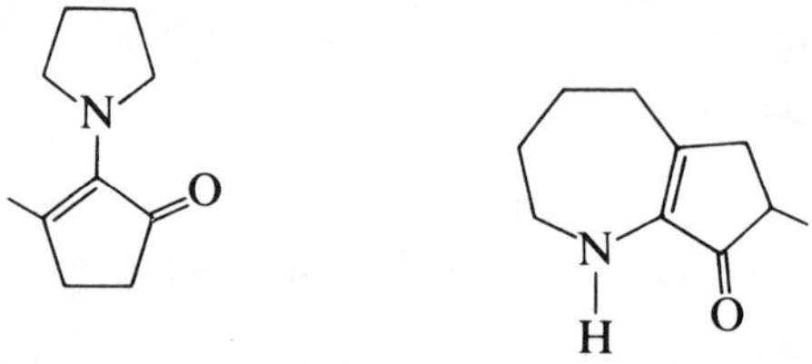

FIG. 11. Two bitter compounds from proline–monosaccharide or proline–cyclotene interaction: left, bitter and astringent (threshold 50 ppm); right, bitter (threshold 10 ppm).

dimethyl-2-cyclopenten-1-one (4-methylcyclotene) reaction product is even more bitter. Such compounds probably contribute to the bitterness of roasted foods.

HYDROXYPROLINE

Whereas proline gave rise to more than 120 proline-specific compounds, some of which have bready aroma and some of which have a bitter taste, hydroxyproline gave less than 40 hydroxyproline-specific compounds.[19] Most of the compounds identified were pyrrole rather than pyrrolidine derivatives. No bready aroma or bitter taste was found. The compounds of most interest are listed in Fig. 12, together with their aromas and thresholds.

The last compound in Fig. 12 is clearly formed via 2,3-enolisation, whereas the two preceding ones are derived by 1,2-enolisation.

Two indolizinones were identified amongst the products of the hydroxyproline–monosaccharide reaction. The structures of these bi-

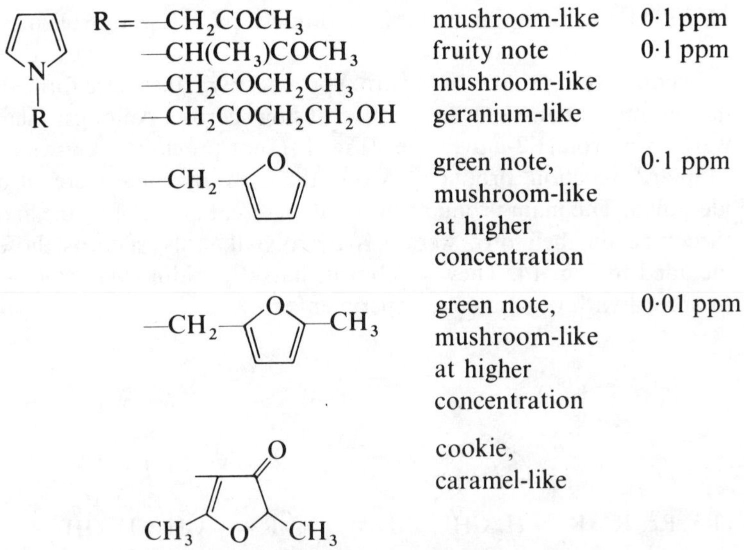

FIG. 12. Some *N*-substituted pyrroles from hydroxyproline–glucose and/or hydroxyproline–rhamnose interaction, with their associated aromas and thresholds (in water).

cyclic compounds are given in Fig. 13. It is interesting to note that, in contrast to the pyrrolizines formed from proline, in the indolizines formed from hydroxyproline the aromatic ring is derived from the amino acid and the hydrogenated ring from the carbohydrate.

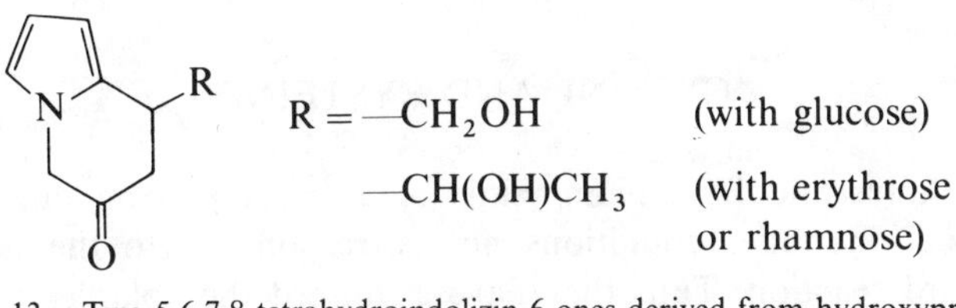

FIG. 13. Two 5,6,7,8-tetrahydroindolizin-6-ones derived from hydroxyproline.

SERINE AND THREONINE

Baltes and Bochmann[20] have examined the Maillard reactions undergone by these two hydroxyamino acids under three types of conditions: cooking at 100°C in buffer, autoclaving at 120–150°C in buffer, and roasting at 220°C. More than 350 compounds were identified, including pyrazines, pyridines, pyrroles, pyrrolizines, furans, furanones and oxazoles.

In contrast to proline and hydroxyproline, pyrazines were formed; in fact, more than 120 were obtained using sucrose. Amongst them were 15 pyrrolo[1,2-*a*]pyrazines (Fig. 14), not previously known as Maillard reaction products. More than 70 pyrroles were also identified. The main products of pyrolytic breakdown of serine and threonine on their own were 9 *N*-pyrrolylalkanols, such as those included in Fig. 14. They had burnt, harsh, pyridine-like aromas, in accord with organoleptic experiments.[21]

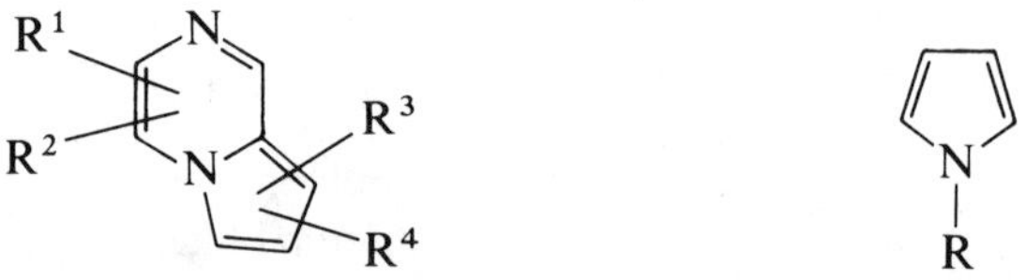

$R^1, R^2, R^3, R^4 = H, CH_3, C_2H_5$

$R = —CH_2CH_2OH$
$—CH_2CH(OH)CH_3$
$—CH_2CH_2CH_2OH$

FIG. 14. Left: pyrrolopyrazines from serine– and threonine–sucrose interaction. Right: *N*-pyrrolylalkanols from pyrolysis of serine and threonine.

Baltes and Bochmann found that pyrolysis of these hydroxyamino acids can produce many of the compounds heretofore regarded as Maillard reaction products, e.g. pyrazines, pyrroles and pyridines.

CYSTINE AND CYSTEINE

Like Baltes and Bochmann, Shu *et al.*[22] have been looking at degradation under conditions not corresponding to the normal Maillard reaction. Thus they have examined the volatiles formed from dilute aqueous solutions of cystine at 160°C, 30 min. At pH 5·5, a complex mixture with an odour described as oniony, vegetable, cabbage and garlic was formed, from which 38 compounds were identified. Thiazoles, sulphides, thiolanes, and thianes dominated. No pyrazines were found.

At pH 2·3, trithiolanes formed more readily, the mixture was less complex, and its odour described as heated, roasted, oniony and biting.

Cysteine solutions were examined in a parallel way.[23] At pH 2·1, 1,2,3-trithia-5-cycloheptene (Fig. 15) dominated, with 2-thiophenethiol as the second major component. At pH 5·1, the yield was four times greater, but this mixture was less complex, acetone and the 3,5-dimethyl-1,2,4-trithiolanes being the main components. At pH 7·1, the yield dropped to one-tenth.

Shu *et al.*[24] also examined the degradation of furaneol under these conditions and found that it gave mainly acetone at pH 2·2, but 2,3-pentanedione at pH 5·1, which remained the largest single product at pH 7·1. There was evidence of the formation of new C–C bonds, e.g. production of 2,4,5-trimethyl-3(2*H*)-furanone.

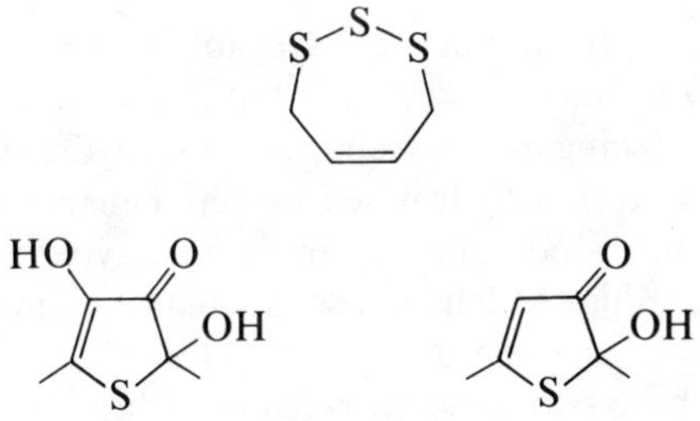

FIG. 15. Above: trithiacycloheptene from aqueous cysteine at pH 2·1, 160°C, 30 min. Below: two thiophenones from aqueous cysteine–furaneol at pH 2·4, 160°C, 30 min.

On treating a mixture of cystine and furaneol similarly, the same group[25] obtained two interesting new compounds, 2,5-dimethyl-2,4-dihydroxy-3(2*H*)-thiophenone and 2,5-dimethyl-2-hydroxy-3(2*H*)-thiophenone, which together with 2,5-dimethyl-4-hydroxy-3(2*H*)-thiophenone and 2,4-hexanedione accounted for almost 40% of the volatiles. The first compound had pot-roasted character in both aroma and taste; the second was described as roasted onion in aroma and taste.

Hartman and Ho[26] examined the volatiles produced from the reaction of the sulphur amino acids cysteine, cystine and methionine with butanedione, a Maillard intermediate, in refluxing diglyme or glycerol at 160°C. Several heterocyclic compounds were formed in each mixture, that from cystine being the most complex. Methionine gave no thiazoles, but 2-(4-methylthiobutyl)-3,5,6-trimethylpyrazine, and methional and its aldol condensation product.

PHENYLALANINE

Refluxing in aqueous buffer with 6 moles of glucose led to a range of furans, some pyrans, only two pyrroles, cyclotene, 2,3-pentanedione, 3-hydroxy-2-butanone, and several compounds with benzene rings: benzaldehyde, phenylacetaldehyde, phenol, benzyl alcohol, 1-phenyl-2-propanone, 2(?)-phenyl-2(?)-propenal, and 2-phenyl-2-butenal.[27] These compounds too provide evidence both of aldol condensation and of dealdolisation (see earlier).

LEUCINE

A recent patent[28] for a chicken flavorant is based on a Maillard reaction involving leucine primarily, but also a sulphur-containing compound, a reducing monosaccharide and a hydroxylated solvent. Serine, chicken fat and hydrolysed protein enhanced the product. The sulphur compounds quoted are cystine, cysteine, sodium sulphide and glutathione. Rhamnose as reducing monosaccharide enhanced the flavour intensity.

This patent brings to mind work cited earlier,[13,14] but also two questions in particular:

(a) Do peptides differ in their action as compared with their constituent amino acids?

(b) What is the function of fat?

As regards the first, all too little is known. Merritt and Robertson[30] submitted several dipeptides to pyrolysis in 1967. Van Chuyen *et al.*[31] showed that dipeptides browned much more (about ten-fold) with glyoxal at 100°C, pH 5·0, than the corresponding amino acids. Acetaldehyde, though less reactive, also led to marked differences. With equimolar amounts of glycine and its oligomers up to the tetrapeptide and glucose, the absorbances at 470 nm were 0·003, 0·055, 0·045 and 0·090, respectively, after 9 hours at 100°C. Pyrazinones, formed from ten dipeptides on heating with glyoxal, were isolated and characterised. Whereas glycyl-leucine is very bitter, the pyrazinone derived from it is astringent and a little sour. The reaction of glycyl-alanine with diacetyl and with pyruvaldehyde has been briefly examined.[32] Overall, many aspects remain to be explored.

As regards the second question, fat clearly has a physicochemical function in providing a phase into which many flavorants partition preferentially, but it is also a source of reactants, particularly as a result of oxidation. A recent paper by Saittagaroon *et al.*[33] on roasted coconut is relevant. Unroasted coconut provides lactones as primary odorants, which mostly survive roasting, when pyrazines in particular are added to them. If defatted coconut is roasted, there is a great increase in the strong nut-like and burnt odour, the deficiency in sweet aroma being attributed to the absence of δ-lactones and esters, i.e. the normal odour of roasted coconut is dominated by lipid-derived volatiles, not unlike the situation in roast beef.[34]

Raising the question of peptides leads to comparison between mono- and di-saccharides also. According to Spark's results, browning of glycine with sucrose and maltol in aqueous solution at 50° is very much slower than with monosaccharides.[35] However, Ledl *et al.*[36] focused on the 1-deoxy-2,3-dicarbonyl intermediates. The cyclic isomer (see Fig. 16), when R = H, can dehydrate to the γ-pyrone,

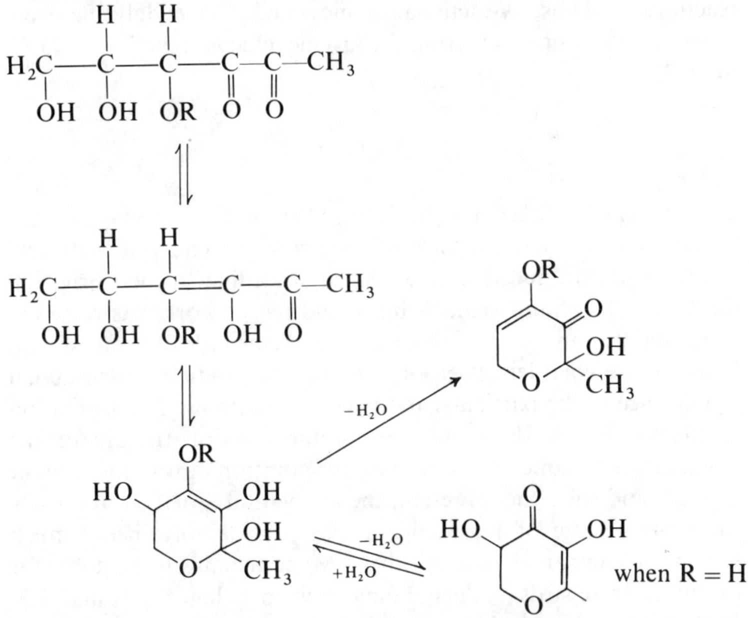

FIG. 16. The effect of 4-*O*-substitution on the dehydration of the 1-deoxy-2,3-dicarbonyl intermediate.

dihydrohydroxymaltol, but this is not possible when it is derived from a 1,4-disaccharide ($R = C_6H_{11}O_5$); a β-pyrone is formed instead. The β-pyrone ($R = \beta$-galactosyl) has been isolated from a lactose–glycine reaction mixture and detected in heated milk,[37] and this isomer ($R = \alpha$-glucosyl) has been identified in ginseng.[38] Both compounds change readily, the former to galactosylisomaltol, the latter to maltol. Maltol and isomaltol are thus typical of disaccharide degradation, though it should be noted that Mills and Hodge[39] did obtain a trace of maltol, as well as 50% dihydrohydroxymaltol and 1% hydroxymaltol, by decomposition of 1-deoxy-1-prolinofructose. However, heating dihydrohydroxymaltol with proline in ethylene glycol at 140°C produced no maltol[39] so, unless positive evidence is available, schemes showing this reaction (e.g. refs 3, 4, 40) should be withdrawn.

GLYCINE

Recently it has been shown that amides are formed in Maillard reactions.[41] Thus, *N*-methylacetamide and *N*-acetylglycine were found as major products from the glycine–glucose reaction at 95°C, pH 6·7.

CONCLUSION

This survey of aroma compounds from the Maillard reaction started by outlining the current view of that complex chemical network. The volatiles produced were classified according to their origin in the sugar, the amino acid, or both, and recent work was reviewed using the amino acid involved as the theme. Ideally one should know all the volatiles of sensory significance produced, their actual significance in the particular food system, and how their formation is affected by all the conditions of the reaction (nature of the reactants, their concentration, the concentration of water and other solvents and substances present, the temperature, pH, etc.). Clearly the current state of knowledge varies considerably, being much greater in some areas than in others. Much remains to be done, for example, as regards understanding how to balance 1,2- and 2,3-enolisation of Amadori compounds and aldolisation and dealdolisation, the effects of oligomers vis-à-vis monosaccharides and amino acids, and all the many interactions which can potentially take place in foodstuffs. However, considerable progress is being made.

ACKNOWLEDGEMENT

Grateful acknowledgement is made of the help with the literature provided by Mrs Beryl Alvarez of Bush, Boake, Allen Ltd.

REFERENCES

1. HODGE, J. E. (1953). *J. Agric. Food Chem.*, **1**, 928–43.
2. NURSTEN, H. E. (1981). Ref. 8, 491–6.
3. TRESSL, R., KOSSA, T. and HOLZER, M. (1975). In: *Geruch- und Geschmackstoffe*, Ed. Drawert, F., Verlag Hans Carl, Nürnberg, 33–47.
4. NURSTEN, H. E. (1980–1). *Food Chem.*, **6**, 263–77.
5. HURRELL, R. F. (1982). In: *Food Flavours. A: Introduction*, Eds Morton, I. D. and Macleod, A. J., Elsevier, Amsterdam, 399–437.
6. VERNIN, G. and PARKÁNYI, C. (1982). In: *Chemistry of Heterocyclic Compounds in Flavours and Aromas*, Ed. Vernin, G., Ellis Horwood, Chichester, 151–207.
7. FUJIMAKI, M., NAMIKI, M. and KATO, H. (Eds) (1986). *Amino-Carbonyl Reactions in Food and Biological Systems*, Elsevier, Amsterdam.
8. ERIKSSON, C. (Ed.) (1981). *Maillard Reactions in Food: Chemical, Physiological and Technological Aspects* (*Progress in Food and Nutrition Science*, Vol. 5), Pergamon Press, Oxford.
9. WALLER, G. R. and FEATHER, M. S. (Eds) (1983). *The Maillard Reaction in Foods and Nutrition*, ACS Symposium Series 215, American Chemical Society, Washington, DC.
10. TRESSL, R., HELAK, B., MARTIN, N. and REWICKI, D. (1986). Ref. 7, 235–44.
11. TRESSL, R., REWICKI, D., HELAK, B. and KAMPERSCHRÖER, H. (1985). *J. Agric. Food Chem.*, **33**, 924–8.
12. SUYAMA, K. and ADACHI, S. (1980). *J. Agric. Food Chem.*, **28**, 546–9.
13. HARTMAN, G. J., CARLIN, J. T., HWANG, S.-S., BAO, Y., TANG, J. and HO, C.-T. (1984). *J. Food Sci.*, **49**, 1398–1400.
14. SHU, C.-K., MOOKHERJEE, B. D., BONDAROVICH, H. A. and HAGEDORN, M. L. (1985). *J. Agric. Food Chem.*, **33**, 130–2.
15. PABST, H. M. E., LEDL, F. and BELITZ, H.-D. (1984). *Z. Lebensm. Unters. Forsch.*, **178**, 356–60.
16. SHIGEMATSU, H., SHIBATA, S., KURATE, T., KATO, H. and FUJIMAKI, M. (1975). *J. Agric. Food Chem.*, **23**, 233–7.
17. TRESSL, R., REWICKI, D., HELAK, B. M., KAMPERSCHRÖER, H. and MARTIN, N. (1985). *J. Agric. Food Chem.*, **33**, 919–23.

18. Tressl, R., Helak, B., Köppler, H. and Rewicki, D. (1985). *J. Agric. Food Chem.*, **33**, 1132–7.
19. Tressl, R., Grünewald, K. G., Kersten, E. and Rewicki, D. (1985). *J. Agric. Food Chem.*, **33**, 1137–42.
20. Baltes, W. and Bochmann, G. (1986). Ref. 7, 245–55.
21. Lane, M. J. and Nursten, H. E. (1983). Ref. 9, 141–58.
22. Shu, C.-K., Hagedorn, M. L. and Mookherjee, B. D. (1985). *J. Agric. Food Chem.*, **33**, 438–42.
23. Shu, C.-K., Hagedorn, M. L., Mookherjee, B. D. and Ho, C.-T. (1985). *J. Agric. Food Chem.*, **33**, 442–6.
24. Shu, C.-K., Mookherjee, B. D. and Ho, C.-T. (1985). *J. Agric. Food Chem.*, **33**, 446–8.
25. Shu, C.-K., Hagedorn, M. L., Mookherjee, B. D. and Ho, C.-T. (1985). *J. Agric. Food Chem.*, **33**, 638–41.
26. Hartman, G. J. and Ho, C.-T. (1984). *Lebensm. Wiss. Technol.*, **17**, 171–4.
27. Mevissen, L. and Baltes, W. (1983). *Z. Lebensm. Unters. Forsch.*, **176**, 206–7.
28. Hartman, G. J., Scheide, J. D. and Ho, C.-T. (1983–4). *Perf. Flav.*, **8** (Dec./Jan.), 81–6.
29. Tandy, J. S. (1985). Eur. Pat. 0 160 794.
30. Merritt, C., Jr and Robertson, D. H. (1967). *J. Gas Chromatog.*, **5**, 96–8.
31. Van Chuyen, N., Kurata, T. and Fujimaki, M. (1973). *Agric. Biol. Chem.*, **37**, 327–34.
32. Piloty, M. and Baltes, W. (1979). *Z. Lebensm. Unters. Forsch.*, **168**, 368–73.
33. Saittagaroon, S., Kawakishi, S. and Namiki, M. (1984). *Agric. Biol. Chem.*, **48**, 2301–7.
34. Mottram, D. S and Edwards, R. A. (1983). *J. Sci. Food Agric.*, **34**, 517–22.
35. Spark, A. A. (1969). *J. Sci. Food Agric.*, **20**, 308–16.
36. Ledl, F., Fritsch, G., Hiebl, J., Pachmayr, O. and Severin, T. (1986). Ref. 7, 173–82.
37. Ledl, F., Ellrich, G. and Klostermeyer, H. (1986). *Z. Lebensm. Unters. Forsch.*, in press.
38. Matsuura, H., Hirao, Y., Yoshida, S., Kunihiro, K., Fuwa, T., Kasai, R. and Tanaka, O. (1984). *Chem. Pharm. Bull.*, **32**, 4674–7.
39. Mills, F. D. and Hodge, J. E. (1976). *Carbohydrate Res.*, **51**, 9–21.
40. Sakaguchi, M. (1984). *Koryo*, **39**, 85–97.
41. Hayase, F., Kim, S. B and Kato, H. (1985). *Agric. Biol. Chem.*, **49**, 2337–41.

12

The Scientific and Technological Basis of Meat Flavours

G. MacLeod

Department of Food and Nutritional Sciences, King's College, University of London, UK

ABSTRACT

Apart from taste compounds and flavour enhancers, critical components contributing cooked meat flavour are aroma volatiles deriving mainly from thermally induced reactions of e.g. lipids, proteins, peptides, amino acids, sugars and thiamin. A complex network of intertwining reactions leads to the formation of numerous components, many of which act as aroma modifiers (producing oily/fatty, buttery, caramel, roasted, burnt and sulphurous nuances), whilst some are potent trace character impact components contributing meaty aroma. The reactions of low molecular weight compounds, e.g. hydrogen sulphide, ammonia, thiols, sulphides and carbonyls, appear to be important in generating several species-specific aroma components, e.g. sulphur-substituted furan and thiophen derivatives in cooked beef aroma, and some alkylpyridines and sulphur components in sheep meat aromas. Lipid-derived volatiles can make a significant aroma impact in some species, cooked pork aroma being characterised by a high content of unsaturated acids whilst sheep meat aromas contain highly characteristic and unique branched-chain saturated acids. However, lipid is of lesser consequence in cooked beef aroma and, although some lipid is necessary for a full meaty aroma, the intramuscular phospholipids are sufficient. Simulated meat flavourings have been based on spice blends, meat extracts, flavour enhancers, protein hydrolysates, the thermal processing of non-volatile aroma precursors and the incorporation of specific meat flavour-contributing synthetic chemicals. However, authentic meat flavours have been difficult to reconstruct, largely because of the complexity of the natural flavour and due to our current incomplete knowledge regarding meaty character impact compounds. Furthermore, consumer acceptance of meat analogues also depends on a satisfactory binding of desirable meat flavour to the analogue protein, with consequent sustained release during eating. These goals have not been fully achieved.

INTRODUCTION

Meat is an integral part of the diet for many of us, not only as a good source of protein but also because of its sensory properties, of which flavour is one of the most important. An understanding of the scientific basis of meat flavour is therefore of significant interest in its own right. In addition, a demand has been created for authentic simulated meat flavourings for use in convenience soups, gravies, casseroles and instant meals, and also for flavouring meat substitutes manufactured from plant sources. Undoubtedly, therefore, an appreciation of the technological basis of meat flavours is of considerable economic relevance to certain sectors of our food industry.

An enormous research effort to unravel the scientific and technological basis of meat flavours has been channelled along four main routes. First, attempts have been made to define and determine the relevance of non-volatile components present in raw meat. This approach characterised the early literature especially during the 1960s, but several workers have re-vitalised the concept in recent analyses.[1-4] Secondly, advances in analytical instrumentation enabled the characterisation of many volatile components derived from heated meat, particularly during the 1970s. Thirdly, since an adequate explanation of the sensory properties of cooked meat aromas could not be offered on the basis of volatile components identified from natural meat, a large number of model systems approximating cooked meat have been defined and their volatile components analysed. This creative approach has involved the interaction of natural ingredients as well as of specific compounds known to be generated from naturally occurring precursors, and it epitomises much of the patent literature. Lastly, a few attempts have been made to determine the significance of identified volatile components to cooked meat flavours.

This paper summarises some salient conclusions from both the patent and non-patent literature on the flavour of red meats. In doing so, it is necessary to be highly selective and those who wish for a more comprehensive coverage are referred to some detailed reviews published since 1980.[5-10]

NON-VOLATILE COMPONENTS OF MEAT

Raw meat contains three types of non-volatile components which affect flavour, namely taste compounds, flavour enhancers and aroma precursors which generate volatile components mainly on heating. Some examples of taste-active compounds are summarised in Table 1.[6,11-18]

Although monosodium glutamate (MSG) and the 5′-nucleotides supply the characteristic savoury quality of the 'umami' taste, a more important sensory contribution of both is their flavour enhancing property. Kuninaka has suggested that our sensory receptors for flavour enhancement are independent and sterically different from the basic taste receptors, and an explanation of this enhancement in terms of chemical structure is summarised in Fig. 1.[16]

The most important enhancers in meat are glutamic acid, MSG and inosine-5′-monophosphate (IMP) which is formed during ageing from adenosine-5′-triphosphate (ATP). It is the major nucleotide in dead muscle and, because of this, it is more important as an enhancer in natural meat than the closely related guanosine-5′-monophosphate

TABLE 1
Taste Compounds Present in Meat[6,11-18]

Sweet:	glucose, fructose, ribose, glycine,[a] alanine,[a,b] serine, threonine,[b] lysine,[a,b] proline, hydroxyproline[a]
Salty:	inorganic salts,[a] sodium glutamate, sodium aspartate
Sour:	aspartic acid, glutamic acid,[a,b] histidine, asparagine, succinic acid, lactic acid,[a] pyrrolidone carboxylic acid, *o*-phosphoric acid
Bitter:	creatine,[a,c] creatinine,[a,c] hypoxanthine,[a] anserine, carnosine, other peptides, histidine, arginine, methionine,[a] valine, leucine, isoleucine, phenylalanine,[a] tryptophan, tyrosine[a]
'Umami':	MSG, 5′-IMP,[a,b] 5′-GMP,[a] certain peptides

[a] Main taste contributors according to Solms.[17]
[b] Key taste compounds for beef according to Motono.[12]
[c] Controversy exists over real contribution from creatine and creatinine.[14,18]

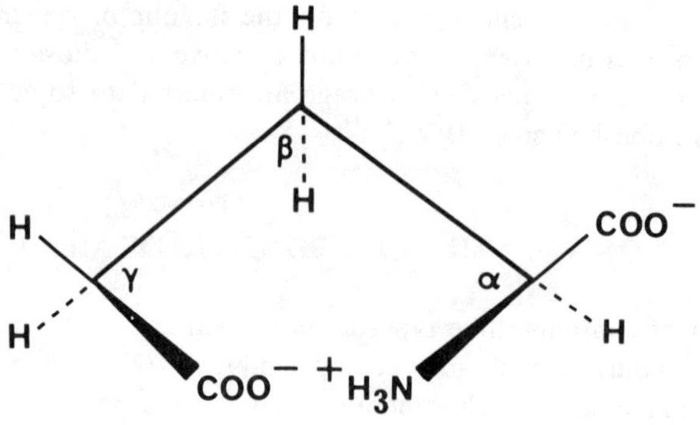

For L-glutamate, the receptor binds the stable five-membered ring formed by electrostatic force between the COO^- and the NH_3^+. Essential features for binding are a planar structure, the α-amino group, the γ-carboxyl group and the α-hydrogen atom.[16]

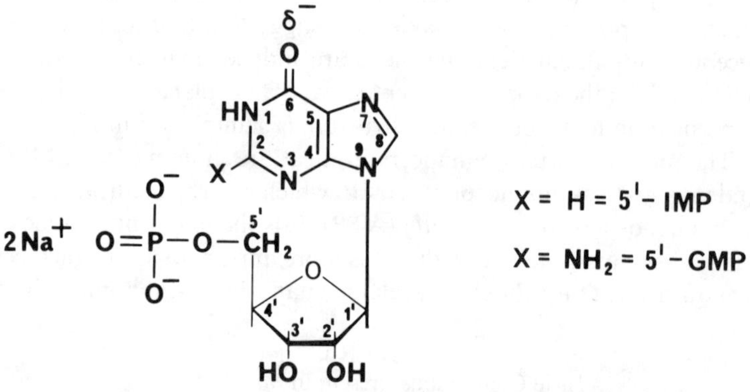

For the 5′-nucleotides, the molecule must bind closely to three receptor sites.[16] Site A binds ionically with the two O^- of the phosphate ester. Site B, which is 8 Å from A, binds with the O (δ^-) on C-6 while site X binds with substituent X on C-2.

FIG. 1. Structural requirements for flavour enhancement.[16]

(GMP), despite the fact that GMP is the more efficient potentiator.[14,16] Both nucleotides exhibit binary synergism when present with glutamate, and a ternary synergism is shown by mixtures of MSG, IMP and certain amino acids, e.g. glycine,[14] or dipeptides, e.g. asparagine-L-aspartate.[9] The 5′-nucleotides enhance meaty, brothy, MSG-like, mouthfilling, dry and astringent qualities; they suppress sulphurous and hydrolysed vegetable protein notes whilst sour, sweet, oily, fatty, starchy and burnt qualities are not affected.[16] This boosting of desirable qualities and inhibition of less pleasing attributes was recently confirmed by Kirova and Peschevska who reported high coefficients correlating the IMP content of beef broths with the sensory parameter 'pleasant, beef broth taste'.[19] In meat during autolysis, IMP is converted slowly into inosine and although this change occurs during chilled storage its effect on perceived flavour is not clear. However, extensive thermal decomposition was recently shown for both IMP and GMP heated at 121°C/30 min in a soup matrix.[20] Such decomposition, during canning at pH 4·5–6·5, involved initial phosphate hydrolysis—converting the nucleotide into the nucleoside—followed by slow base hydrolysis forming hypoxanthine (from IMP) and guanine (from GMP).[20] The nucleosides and bases formed have no flavour enhancing properties.[16] In fact, IMP was more stable than GMP.[20] The same workers concluded that MSG was very stable under these conditions,[20] but contrary evidence appeared in the very same journal issue where Gayte-Sorbier *et al.* reported that, in the pH range 4–6, which is applicable to most foods, native glutamic acid and added MSG are converted into pyrrolidone carboxylic acid (PCA) on storage, boiling and canning, with consequent decreased potentiation and an off-flavour development at certain PCA concentrations.[21]

Since the early 1960s, many workers have reported that the non-volatile precursors of cooked meat aroma are water-soluble components which can be extracted from raw muscle. The compositions of the fractions isolated vary and have been summarised.[3,6] Essentially, the most important aroma precursors include a range of amino acids, peptides, reducing sugars (especially glucose, fructose and ribose) and nucleotides (especially IMP). During ageing, increased levels occur for free amino acids (from enzymic proteolysis), for ribose and ribose-5-phosphate (from ATP and IMP decomposition) and for glucose and fructose (from glycolysis). A pool of Maillard reactants is therefore present before cooking. Furthermore, during the early stages of heating some continued enzymic proteolysis enhances the free amino acid concentration[22] and thermally induced release of amino acids from muscle proteins is also likely throughout cooking.[23] In an interesting current review of the wide range of muscle proteins now characterised, Pearson *et al.* report that there is little doubt that many of the most important compounds contributing to meat flavour are derived from myofibrillar proteins, which are relatively rich in sulphur amino acids, with lesser amino acid contributions generated from collagen and sarcoplasmic proteins, e.g. myoglobin and haemoglobin.[23]

On the above evidence, therefore, it is not surprising that the Maillard reaction is a keynote basis for meat flavour development. Depletions of amino acids and reducing sugars during cooking have been monitored, the most significant losses occurring for cysteine and ribose.[24,25] This discovery led to the now classic 'reaction product' meat flavour patent of Morton *et al.* which involved thermal treatment of cysteine (optionally with other amino acids) and ribose (or other reducing sugars, pentoses being more reactive than hexoses).[25,26]

A final, and worthy, water-soluble non-volatile precursor of cooked meat aroma is thiamin.

Reports on the role of lipids in meat flavour have been confused but current conclusions are as follows. Lipid oxidation starts in raw meat and is catalysed by enzymes in muscle microsomes.[27] It continues during cooking, together with thermal non-oxidative degradations. Even in lean muscle, the intramuscular lipids are a source of many volatile components. So much so, in fact, that they dominate gas chromatograms of cooked lean meat aromas and hinder the identification of more important trace components. The addition of adipose tissue to lean meat does not give rise to a proportional increase of lipid-derived volatiles, showing that the intramuscular marbling fat (mainly triglycerides) and the intramuscular structural or membrane lipids (mainly phospholipids) are the major sources of volatile components.[28,29] These tissue lipids (and largely the phospholipids) have a relatively higher content of unsaturated fatty acids than the depot fats, especially 18:1, 18:2, 18:3 and 20:4 acids which are readily oxidised.

Additionally, and contrary to some early hypotheses, Mottram has shown that species-specific flavour precursors are present in lean meat (beef and pork),[30] although the addition of fat induces fat/lean interactions which enhance species differences.[30] Furthermore, selective removal of triglycerides from lean beef causes no significant difference in the cooked aroma (or in the volatiles liberated during cooking) when compared with a control sample, whereas removal of both triglycerides and phospholipids from lean beef generated marked sensory (less meaty, more roasted) and chemical differences.[1]

It is therefore concluded that some lipid is necessary for full meaty aroma, but the intramuscular tissue phospholipids, which comprise only about 1% of the muscle composition, are sufficient and the triglycerides are not essential.[1] Interactions between water-soluble and phospholipid-derived precursors are probably important. Interestingly, however, lamb fat is more significantly species-specific than either beef or pork fat[31] and, although a number of unique branched-chain saturated acids biosynthesised by the sheep (in particular 4-methyloctanoic acid and 4-methylnonanoic acid) contribute to this ovine specific flavour,[8,32] additional compounds are necessary for a full, readily distinguishable lamb flavour.[31]

Another relevance of tissue lipids, as precursors, is their role as primary contributors to the flavour defect called 'warmed over flavour' (WOF) in cooked meats.[2,8] This is attributed to oxidation of tissue phospholipids and phosphatidylethanolamine plays an important role.[2] This is in contrast to oxidation on frozen storage when triglycerides are mainly involved. A major catalyst in WOF development is non-haem iron released from haem pigments during heating, and evidence suggests that the anti-oxidative activity of Maillard reaction products[33] in overheated meats protects against WOF. Other possible inhibitors include phosphates, nitrite, salt and ascorbate.[2]

VOLATILE COMPONENTS

Gross reactions occurring in meat during cooking are of three main types:

(a) lipid oxidation/degradation;
(b) thermal decomposition and interactions of peptides, amino acids, sugars and ribonucleotides;
(c) thermal degradation of thiamin.

However, reaction products have a habit of becoming reactants, so the end result is a complex and ill-defined network of intertwining reactions. As a consequence, an impressive record of about 750 aroma components have been identified from heated beef,[6,34–40] although less than half this number have been reported from cooked pork and from cooked lamb/mutton aromas,[34] probably because of significantly fewer publications on these two species. A rough breakdown of the chemical classes represented in cooked beef aroma is shown in Table 2. The probability is that major components of the aroma are now established, but some trace components have eluded identification.

The relative sensory significance of these volatiles is not clear but four propositions are made. First, many are likely to be unimportant. Secondly, the premise that the odour quality 'meaty' can be explained by the simultaneous combined effects on the olfactory epithelium of most of the currently identified compounds is now highly unlikely. Thirdly, therefore, it is suggested that meatiness should be cleanly dissected sensorially into about ten odour qualities, in which case many of the identified volatiles act as 'aroma modifiers' contributing buttery, caramel, roasted, burnt, sulphurous, green, fragrant, oily/

TABLE 2
Chemical Classes of Volatile Compounds Reported in Aromas from Heated Beef[6,34–40]

Class	*Number*
Hydrocarbons – aliphatic	73
– alicyclic	4
Terpenoids	8
Alcohols – aliphatic	46
– alicyclic	1
Aldehydes – aliphatic	55
Ketones – aliphatic	44
– alicyclic	8
Carboxylic acids – aliphatic	20
Lactones	32
Esters – aliphatic	27
Ethers – aliphatic	5
Amines – aliphatic	20
Chlorinated compounds	10
Benzenoids	86
Sulphur compounds (not heterocyclic)	68
Furans and derivatives	43
Thiophens and derivatives	40
Pyrroles and derivatives	20
Pyridines and derivatives	17
Pyrazines and derivatives	54
Oxazol(in)es etc.	13
Thiazol(in)es	29
Other sulphur heterocyclics	13
Miscellaneous	12

fatty and nutty notes. Structure/activity correlations (or at least associations) do exist in the literature for many of these odour qualities. Fourthly, some aroma components do contribute a truly specific meaty odour and are therefore character impact compounds. There is no known single unique character impact compound for cooked beef (or pork or lamb/mutton) aroma, although many workers believe that several potent key meaty compounds are present in the natural cooked aroma, many of which remain to be identified. The majority are relatively high-boiling compounds. Clearly, the scientific and technological basis of meat flavours depends primarily on defining such compounds and the reactions which generate them.

A search of the literature for individual chemical compounds described as meaty is presented in Fig. 2.[5,6,40–75] A specific attempt has been made to keep the list relatively 'pure'—sensorially speaking—thereby excluding components possessing very diffuse, composite and poorly defined effects, albeit that they are reported to impart a savoury and somewhat meat-like character, often in conjunction with other components. A prolific number of this latter type of compound abounds in the patent literature. The meaty quality of a few of the compounds entered in Fig. 2 (e.g. **5**, **11**, **32**, **52**, **67**, **68**) has been under attack by some workers, probably because it is a difficult quality to define with precision. Nevertheless, in the absence of a supreme arbitrator, they have been included for completeness. Also, many compounds are meaty only at certain concentrations, usually very low concentrations which are often unspecified. Twenty-four of the compounds presented in Fig. 2 (see footnotes) have been identified in natural cooked meat aromas. Some of these are disputed, as mentioned above; others possess complex multi-quality sensory properties, e.g. the oxazol(in)es and thiazol(in)es, but some, e.g. 2-methyl-3-(methylthio)furan (**15**), have been proposed as true character impact compounds.[40]

Some reactions which generate the various meaty compounds of Fig. 2 are summarised in Table 3 and discussed below. Hydrogen sulphide is of major importance as a reactant. It is liberated in natural meat from cysteine, cystine, methionine, glutathione (γ-Glu-Cyst-Gly) and from thiamin during heating. Rapid evolution occurs from glutathione during the early stages of cooking,[55] but cysteine is the main H_2S precursor on prolonged heating.[76]

The thermal degradation of cysteine in water at 160°C/30 min

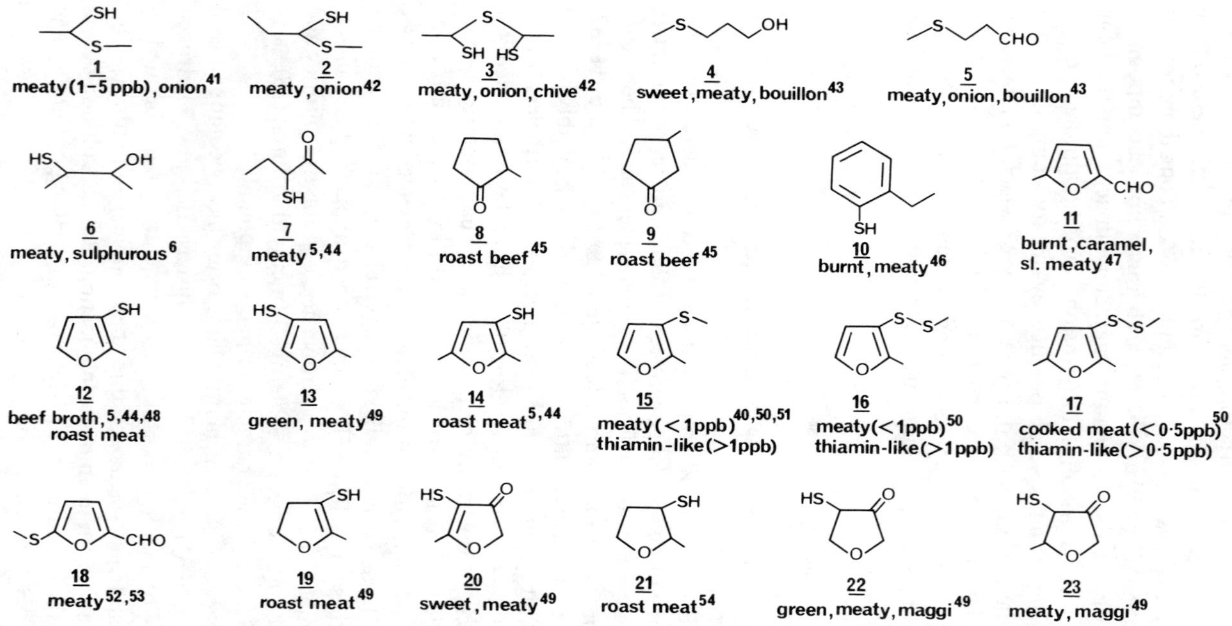

FIG. 2. Some individual chemical compounds described in the literature as meaty.[5,6,40–75] 1. Compounds **1**, **5**, **8**, **9**, **11**, **15**, **18**, **23**, **32**, **43**, **46**, **48–50**, **52**, **53**, **55**, **63**, **64** and **66–68** have been identified in natural cooked beef aroma.[6,34–40] 2. Compounds **5**, **11**, **32**, **46**, **47**, **52**, **66** and **68** have been identified in natural cooked pork aroma.[34] 3. Compounds **3**, **11**, **46**, **55** and **68** have been identified in natural cooked lamb/mutton aroma.[34]

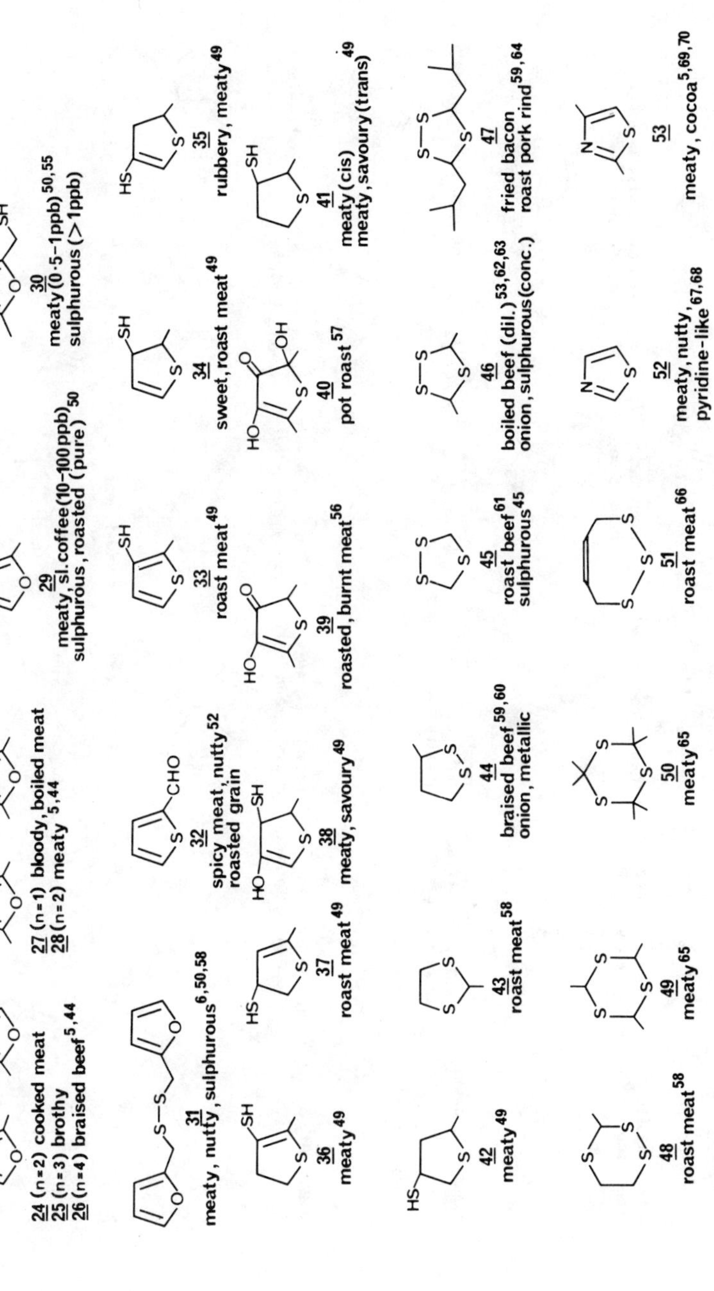
24 (n=2) cooked meat
25 (n=3) brothy
26 (n=4) braised beef 5,44
27 (n=1) bloody, boiled meat
28 (n=2) meaty 5,44
29 meaty, sl. coffee (10–100ppb)
sulphurous, roasted (pure) 50
30 meaty (0·5–1ppb) 50,55
sulphurous (>1ppb)
31 meaty, nutty, sulphurous 6,50,58
32 spicy meat, nutty 52
roasted grain
33 roast meat 49
34 sweet, roast meat 49
35 rubbery, meaty 49
36 meaty 49
37 roast meat 49
38 meaty, savoury 49
39 roasted, burnt meat 56
40 pot roast 57
41 meaty (cis)
meaty, savoury (trans) 49
42 meaty 49
43 roast meat 58
44 braised beef 59,60
onion, metallic
45 roast beef 61
sulphurous 45
46 boiled beef (dil.) 53,62,63
onion, sulphurous (conc.)
47 fried bacon
roast pork rind 59,64
48 roast meat 58
49 meaty 65
50 meaty 65
51 roast meat 66
52 meaty, nutty,
pyridine-like 67,68
53 meaty, cocoa 5,69,70

FIG. 2.—*contd.*

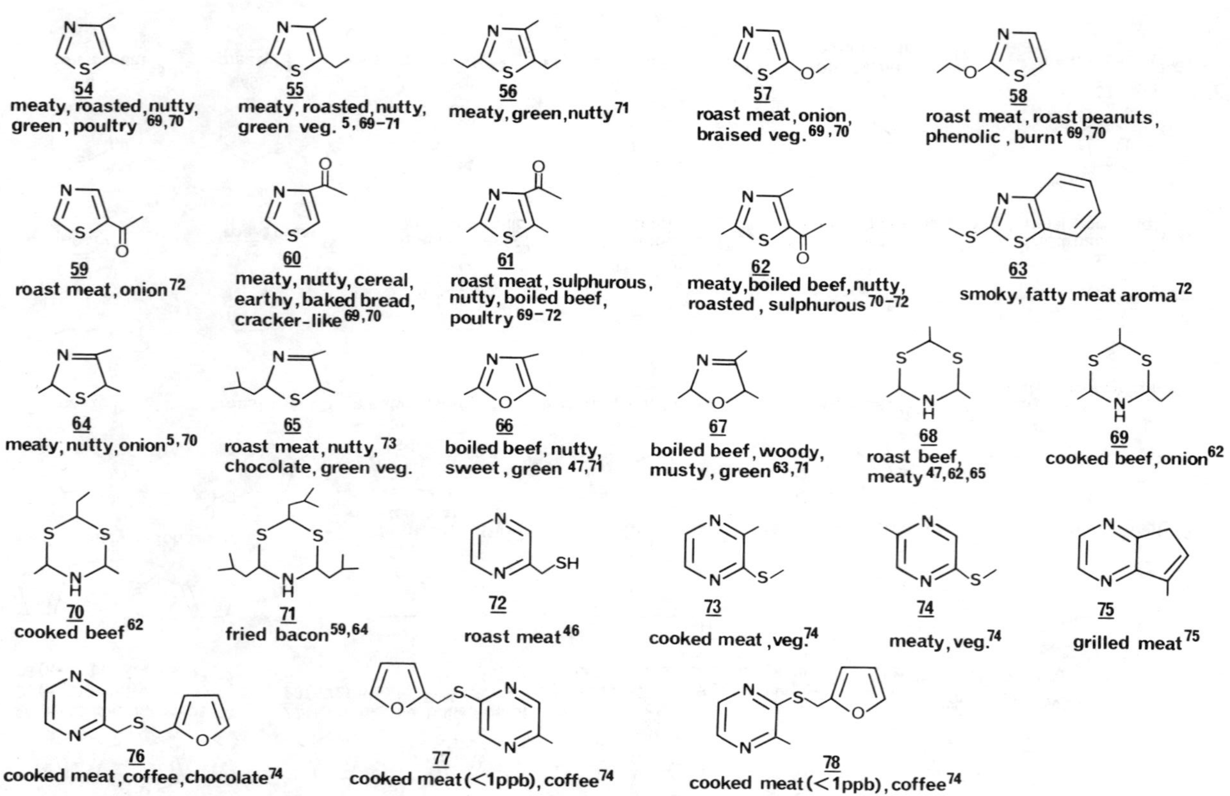

54 meaty, roasted, nutty, green, poultry[69,70]

55 meaty, roasted, nutty, green veg.[5,69–71]

56 meaty, green, nutty[71]

57 roast meat, onion, braised veg.[69,70]

58 roast meat, roast peanuts, phenolic, burnt[69,70]

59 roast meat, onion[72]

60 meaty, nutty, cereal, earthy, baked bread, cracker-like[69,70]

61 roast meat, sulphurous, nutty, boiled beef, poultry[69–72]

62 meaty, boiled beef, nutty, roasted, sulphurous[70–72]

63 smoky, fatty meat aroma[72]

64 meaty, nutty, onion[5,70]

65 roast meat, nutty, chocolate, green veg.[73]

66 boiled beef, nutty, sweet, green[47,71]

67 boiled beef, woody, musty, green[63,71]

68 roast beef, meaty[47,62,65]

69 cooked beef, onion[62]

70 cooked beef[62]

71 fried bacon[59,64]

72 roast meat[46]

73 cooked meat, veg.[74]

74 meaty, veg.[74]

75 grilled meat[75]

76 cooked meat, coffee, chocolate[74]

77 cooked meat (<1ppb), coffee[74]

78 cooked meat (<1ppb), coffee[74]

FIG. 2.—*contd.*

TABLE 3
Some Reactions Generating the Meaty Compounds of Fig. 2

Reaction	*Compound*
Thermal degradation of cysteine[66]	**43, 44, 46, 48, 51, 52**
Thermal degradation of cystine[83]	**9, 43, 44, 46, 48, 49, 52, 59**
Cysteine/pyruvaldehyde[77]	**52**
Cysteine/butanedione[78]	**66**
Cysteine/rhamnose[58]	**11, 14, 30, 46, 43, 48**
Cysteine/methionine/2-furaldehyde[58]	**31**
Cysteine/methionine/ 5-methyl-2-furaldehyde[50]	**30**
Cysteine/glucose	**32**,[81] **48**,[80] **68**[80]
Cysteine/cystine-ribose[81]	**48, 55**
Cyclotene/H_2S/NH_3 [45]	**8, 9, 45**
Acetaldehyde/H_2S[84]	**3, 46, 49**
Acetaldehyde/H_2S/NH_3 [84]	**68**
Acetaldehyde/CH_3SH/H_2S[84]	**1**
Propanal/CH_3SH/H_2S	**2**
Propanal/acetaldehyde/CH_3SH/H_2S	**69, 70**
3-Methylbutanal/H_2S/NH_3 [64]	**47, 71**
Acetaldehyde/pyruvaldehyde/H_2S/NH_3 [84]	**53**
Acetaldehyde/acetoin/NH_3 [71]	**67**
Crotonal/H_2S[84]	**44**
Butanedione/H_2S	**6**
Pentane-2,3-dione/H_2S[5,84]	**7**
Butanedione/acetaldehyde/NH_3 [5,84]	**67**
Butanedione/acetaldehyde/NH_3/H_2S [5,84]	**64, 66**
Pentane-2,3-dione/acetaldehyde/NH_3/H_2S[84]	**55**
HMFone/H_2S[49,86,87]	**13, 19, 20, 22, 23, 33, 34, 35, 36, 37, 38, 41, 42**
HDFone/H_2S[86]	**14, 39**
HDFone/cystine[57]	**7, 39, 40, 43, 46, 48, 52**
Thermal degradation of thiamin	**7**,[48,107] **12**,[48,107,108] **14**,[107] **19**,[48,107] **21**,[107] **24**,[107,108] **54**,[48,107]
Thiamin/CH_3SH	**15, 16, 25, 26**

and at pH 2·2, 5·1 and 7·1 was reported recently by Shu *et al.*[66] They identified twenty-six volatile products including 2-methyl-1,3-dithiolane (**43**), 3-methyl-1,2-dithiolane (**44**), 3,5-dimethyl-1,2,4-trithiolane (**46**), 3-methyl-1,2,4-trithiane (**48**), 1,2,3-trithiacyclohept-5-ene (**51**) and thiazole (**52**). The most vigorous reaction occurred at the IEP of 5·1. The volatile mixture was simple and contained 55% acetone. This pH also favoured the formation of the isomeric 3,5-dimethyl-1,2,4-trithiolanes (**46**). At pH 2·2, the main volatile

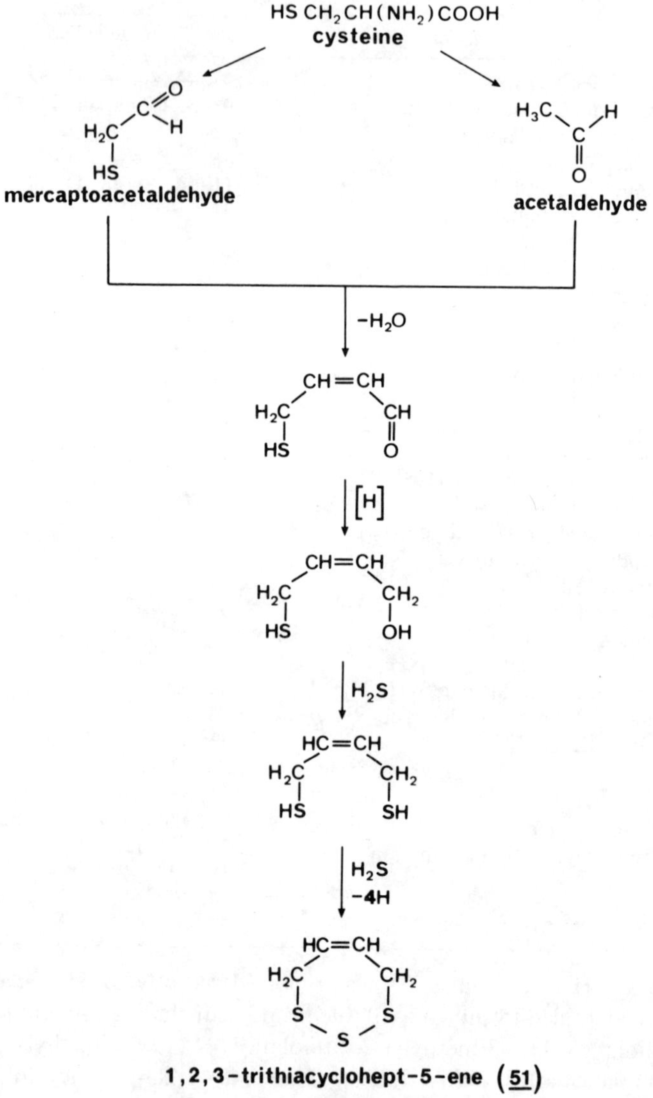

FIG. 3. Formation of 1,2,3-trithiacyclohept-5-ene (**51**) from mercaptoacetaldehyde and acetaldehyde.[66]

products were cyclic sulphur compounds containing 5-, 6- and 7-membered rings, together with thiophens. The major component (34% of the isolate) was the novel meaty compound 1,2,3-trithiacyclohept-5-ene (**51**). It was suggested that it formed by interaction of mercaptoacetaldehyde and acetaldehyde, as shown in Fig. 3. Only a mild degradation occurred at pH 7·1 and, apart from butanone, only cyclic compounds were identified. Major products were the 3,5-dimethyl-1,2,4-trithiolanes (**46**), 3-methyl-1,2,4-trithiane (**48**) and 2-propanoylthiophen.

During Strecker degradation, cysteine initially gives rise to mercaptoacetaldehyde, acetaldehyde, H_2S and NH_3, all of which undergo further reactions. Figure 2 compounds arising in this way include thiazole (**52**; from cysteine/pyruvaldehyde[77]) and 2,4,5-trimethyloxazole (**66**; from cysteine/butanedione[78]). In the presence of reducing sugars, the number of volatile products generated from cysteine increases enormously. At high temperatures, approximating roasting conditions, the products formed reflect the fact that amino acid pyrolysis (rather than Strecker degradation) is an important reaction[79] and cysteine is transformed into mercaptoacetaldehyde, acetaldehyde, H_2S, cysteamine and ethane-1,2-dithiol, which undergo aldol-type and other condensation reactions with sugar degradation products.[58]

A model system of cysteine/rhamnose under Shigetmatsu conditions generated three primary compounds in approximately equal amounts, i.e. 2,5-dimethyl-4-hydroxy-3(2*H*)-furanone (HDFone), 2,5-dimethyl-3(2*H*)-furanone and 5-methyl-2-furaldehyde (**11**), which were transformed into the corresponding sulphur constituents which possessed meat-like aromas.[58] Figure 2 compounds produced (apart from **11**) were 2,5-dimethylfuran-3-thiol (**14**), 5-methylfurfurylthiol (**30**), 3,5-dimethyl-1,2,4-trithiolane (**46**), 2-methyl-1,3-dithiolane (**43**) and 3-methyl-1,2,4-trithiane (**48**).[58] The likely precursor of the last two meaty compounds is ethane-1,2-dithiol,[58] as shown in Fig. 4. Related model systems of 2-furaldehyde/cysteine/methionine and of 5-methyl-2-furaldehyde/cysteine/methionine gave rise to the meaty bis-(2-furfuryl) disulphide (**31**)[58] and 5-methylfurfurylthiol (**30**) respectively.[50]

Further examples of Fig. 2 compounds arising from Maillard type reactions are thiophen-2-carboxaldehyde (**32**),[80] 3-methyl-1,2,4-trithiane (**48**)[79] and thialdine (**68**)[79] from cysteine/glucose, 3-methyl-1,2,4-trithiane (**48**) and 2,4-dimethyl-5-ethylthiazole (**55**) from cysteine/cystine-ribose,[81] and the 2- and 3-methylcyclopentanones (**8, 9**)

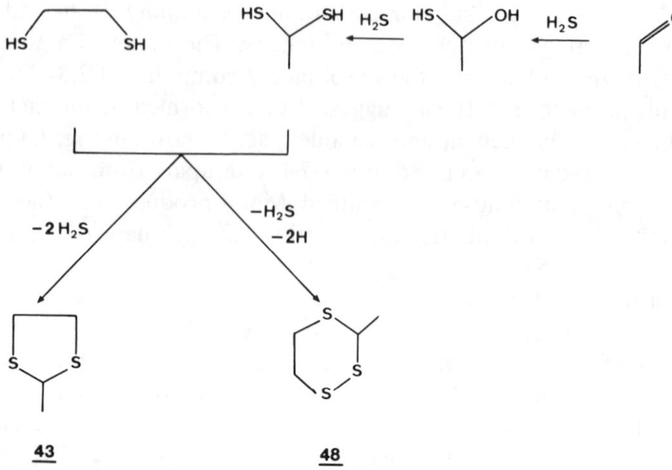

FIG. 4. Formation of 2-methyl-1,3-dithiolane (**43**) and 3-methyl-1,2,4-trithiane (**48**) from ethane-1,2-dithiol/acetaldehyde/H_2S.[58]

and 1,2,4-trithiolane (**45**) from cyclotene/H_2S/NH_3 reaction.[45] The latter compound (**45**) has also been reported from glucose/H_2S.[82]

Shu *et al.* have recently studied the thermal degradation of cystine in water at 160°C/30 min and at pH 5·5 and 2·3 respectively.[83] Forty-two compounds were identified and they were mainly thiazoles, sulphides, thiolanes and thianes. Significantly more thiazoles were present than in the corresponding cysteine system.[66] Figure 2 compounds generated from cystine thermally were 3-methylcyclopentanone (**9**), 2-methyl-1,3-dithiolane (**43**), 3-methyl-1,2-dithiolane (**44**), 3,5-dimethyl-1,2,4-trithiolane (**46**), 3-methyl-1,2,4-trithiane (**48**), 2,4,6-trimethyl-1,3,5-trithiane (**49**, trithioacetaldehyde), thiazole (**52**) and 5-acetylthiazole (**59**).[83] Shu *et al.* reported that, on the whole, cyclic sulphur compounds were generated more readily at pH 2·3, which is the converse of what was reported for the cysteine system where the thiolanes formed preferentially at the IEP rather than below it.[66] It was suggested that this was related to the thermal stability of the S–S bond of cystine at different pH values.

Often an integral part of the degradations just described, but not necessarily so, is a series of reactions of simple low molecular weight compounds, in particular aliphatic aldehydes, alkane-2,3-diones, methanethiol, H_2S and NH_3. These reactions, and the Fig. 2 compounds generated from them, are summarised in Table

3.[5,64,71,84] Acetaldehyde is involved in many of them and its relevant reactions are schematically represented in Fig. 5. Thialdine (**68**; i.e. 2,4,6-trimethyl-1,3,5-dithiazine) is also formed on storage decomposition of 2,4,6-trimethyltetrahydro-1,3,5-thiadiazine, itself formed from bis-(1-mercaptoethyl) sulphide (**3**)/NH_3.[85] 3-Methyl-1,2-dithiolane (**44**) is formed from crotonal/H_2S,[84] 3-hydroxybutane-2-thiol (**6**) from butanedione/H_2S and 3-mercaptopentanone (**7**) from

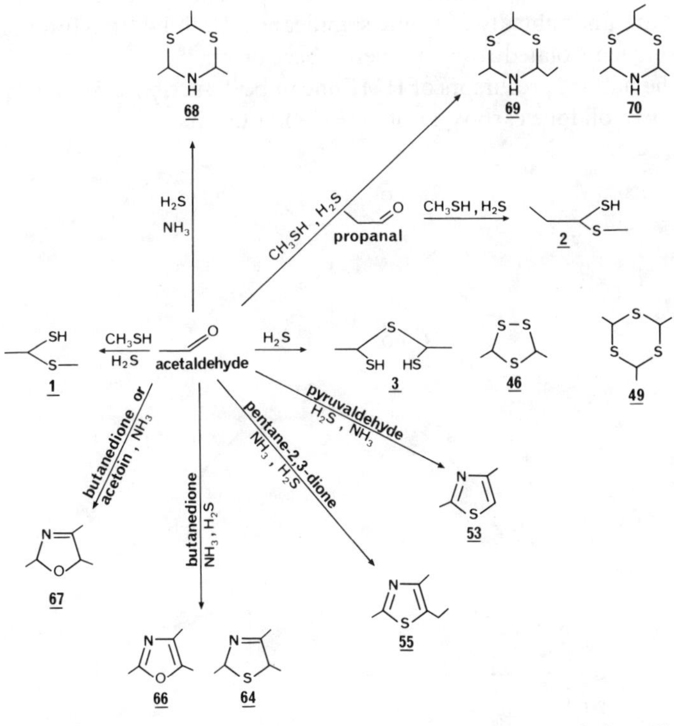

FIG. 5. Schematic reactions of acetaldehyde generating compounds from Fig. 2.

pentane-2,3-dione/H_2S.[5,84] Katz considers that 4- and 5-carbon mercaptoketones are significant in meat flavours and predicts their presence in cooked meats.[5] Further reaction of such compounds with H_2S creates α,β-alkanedithiols, also thought to play an important role in meat flavours.[5] The reaction of 3-methylbutanal/H_2S/NH_3 was reported recently by Shu *et al.*[64] who identified eleven volatile reaction products from a 3-methylbutanal/ammonium sulphide mixture held at only 15°C/2 h. The products included 3,5-di-(2-methylpropyl)-1,2,4-trithiolane (**47**) and 2,4,6-tri-(2-methylpropyl)tetrahydro-1,3,5-dithiazine (**71**) possessing characteristic fried bacon odours.[64] Their formation is sketched in Fig. 6.

The most interesting of all reactions in meat flavour development is the reaction of H_2S with 4-hydroxy-5-methyl-3(2*H*)-furanone (HMFone)[49,86,87] or 2,5-dimethyl-4-hydroxy-3(2*H*)-furanone (HDFone).[86] In both cases, the reaction product mixtures are meaty and are undoubtedly of prime significance. Both hydroxyfuranones have been isolated from simmered beef broth.[88,89]

The natural precursors of HMFone in beef are ribose-5-phosphate and pyrrolidone carboxylic acid (PCA) or taurine (to a lesser extent)

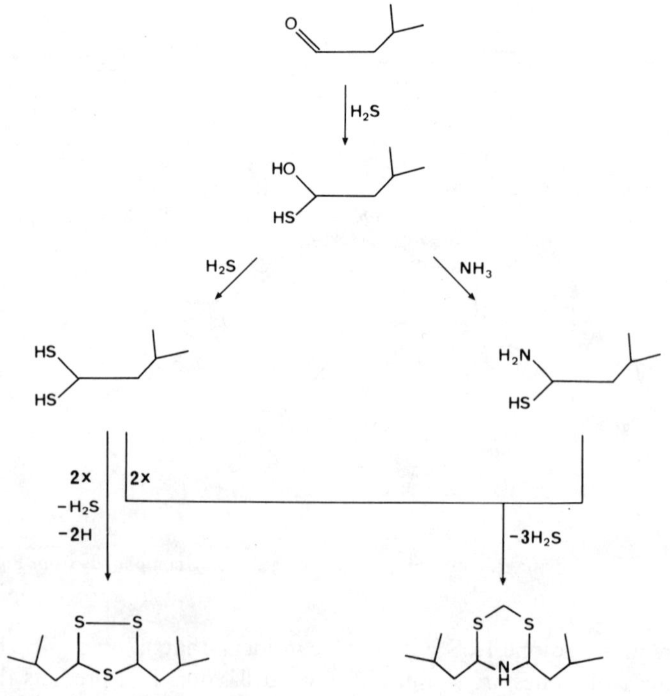

47 **71**

FIG. 6. Formation of 3,5-di-(2-methylpropyl)-1,2,4-trithiolane (**47**) and 2,4,6-tri-(2-methylpropyl)tetrahydro-1,3,5-dithiazine (**71**) from 3-methylbutanal/H_2S/NH_3.[64]

or both.[89] When heated at 100°C/2·5 h in dilute aqueous medium of pH 5·5 they generate HMFone:[89] PCA itself is readily formed on heating NH_3 and glutamic acid or glutamine.[89] Other reactions which generate HMFone include heating ribose-5-phosphate (no amine) in concentrated buffer of pH 5·5,[90] or heating ribose, xylose or glucuronic acid with an amine.[91–93] Hicks and coworkers were the first to produce HMFone from a sugar other than a pentose, and they have suggested that hexuronic acid-containing compounds (e.g. heparin and blood group polysaccharides, as well as certain glycoproteins) are probable sources of HMFone.[91,92] A recent patent describes the formation of HMFone by heating a solution (pH 1–7) of a 5-keto-aldohexonic acid, preferably 5-ketogluconic acid, in a polar solvent at 70–150°C.[94] The furanone can also derive from 5′-ribonucleotides via ribose-5-phosphate obtained, for example, by heating at 60°C,[95] autolysis using aqueous muscle extract[96] or hydrolysis of IMP and/or GMP.[90,97] Yeast extract is therefore a potential contributor and also has a high glutamic acid content.

Precursors of HDFone are C_6 sugars such as glucose or fructose under Maillard conditions via 1-deoxyhexosone,[98] or glucose in the absence of amine[99] or ascorbic acid.[100,101] Takei has reported that ethyl oleate from egg lecithin is important for HDFone production,[102] and further evidence of the useful presence of lipid is given by Doornbos *et al.* who heated Amadori products in a triglyceride at 140°C.[103] However, the instability of HDFone as a solid in air and also in aqueous solution is a problem.[104] Its optimum stability is at pH 4 and its decomposition increases rapidly with temperature.[104] Shu *et al.* recently thermally degraded HDFone in water at 160°C/30 min at pH 2·2, 5·1 and 7·1.[105] Degradation occurred more readily at the lower pH values. They identified twenty volatile products, most of which were reactive simple aliphatic carbonyls and dicarbonyls; the remainder of the volatile reaction product mixture consisted of alkyl substituted 3(2*H*)-furanones which were favoured at higher pH values. Most of these were shown to derive from pentane-2,3-dione as intermediate, indicating that the HDFone ring opened intially on heating.

The reaction of HMFone/H_2S is summarised in Fig. 7. Again ring opening is indicated and an initial partial substitution of the ring oxygen with sulphur occurs. The majority of the reaction products are mercaptofuranoids and mercaptothiophenoids, most

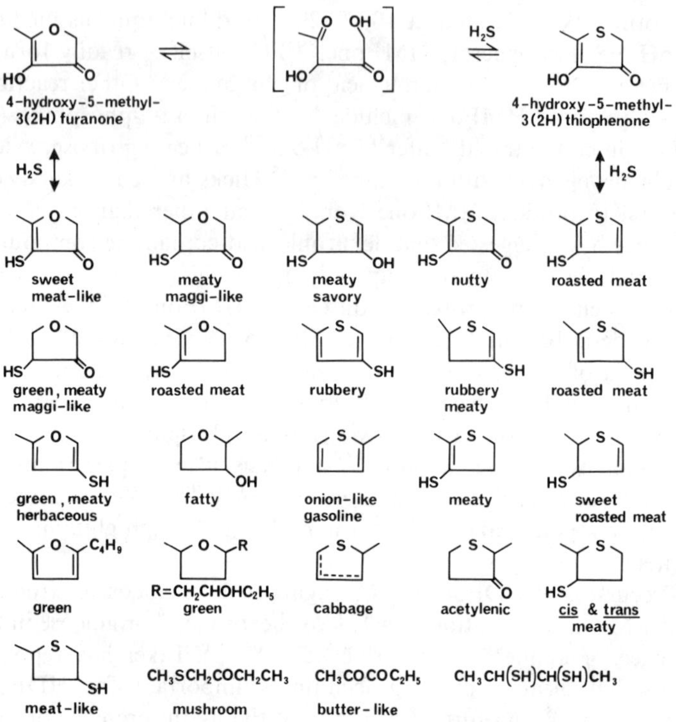

FIG. 7. Reaction of 4-hydroxy-5-methyl-3(2*H*)-furanone (HMFone) and H_2S.[49,86,87]

of which possess meaty odours.[49] The only one identified from meat is 5-methyl-4-mercapto-tetrahydrofuran-3-one (**23**). This was reported in the volatiles obtained on heating (121°C/2 h) defatted, freeze-dried lean beef with the triglyceride fraction (only) extracted from the beef.[35] It is more than likely, however, that the remainder are present in natural cooked beef aroma. Figure 2 compounds identified from heated HMFone/H_2S include 2-methylfuran-4-thiol (**13**), 2-methyl-4,5-dihydrofuran-3-thiol (**19**), 5-methyl-4-mercapto-2,3-dihydrofuran-3-one (**20**), 4-mercapto-tetrahydrofuran-3-one (**22**), 5-methyl-4-mercapto-tetrahydrofuran-3-one (**23**), 2-methylthiophen-3-thiol (**33**), 2-methyl-2,3-dihydrothiophen-3-thiol (**34**), 2-methyl-2,3-dihydrothiophen-4-thiol (**35**), 2-methyl-4,5-dihydrothiophen-3-thiol (**36**), 2-methyl-4,5-dihydrothiophen-4-thiol (**37**), 2-methyl-4-hydroxy-2,3-dihydrothiophen-3-thiol (**38**), 2-methyl-tetrahydrothiophen-3-one (**41**) and 2-methyl-tetrahydrothiophen-4-thiol

(**42**).[49] Bodrero *et al.*, using surface response methodology to study the contribution of various volatiles to cooked beef aroma, showed that the HMFone/H_2S reaction product mixture gave the highest predicted score of their entire study.[106] HDFone reacts with H_2S in a similar manner[49,86,87] and explains two of the Fig. 2 compounds, namely 2,5-dimethylfuran-3-thiol (**14**) and 2,5-dimethyl-4-hydroxy-2,3-dihydrothiophen-3-one (**39**).[86] Dimerisation of **14** could yield **27** and **28**. The chances are that other compounds identified in this reaction are meaty also, but their individual sensory properties have not been reported.

Shu *et al.* recently reacted HDFone with cystine at 160°C/30 min in water (pH 2·4).[57] The volatile products changed on storage for 2 weeks at 4°C and were therefore analysed after holding for 2 weeks. When compared with the products generated from HDFone[105] and from cystine[83] treated similarly, the major difference was the presence of relatively large concentrations of hexane-2,4-dione (16%) and of three thiophenones (22·5%). One of these, i.e. 2,5-dimethyl-4-hydroxy-2,3-dihydrothiophen-3-one, has been reported previously from HDFone/H_2S,[86] but the other two are novel, namely 2,5-dimethyl-2-hydroxy-2,3-dihydrothiophen-3-one (possessing a roasted onion odour) and 2,5-dimethyl-2,4-dihydroxy-2,3-dihydrothiophen-3-one (**40**) possessing a pot roast aroma and taste. Apart from this compound (**40**), other Fig. 2 compounds identified were 3-mercaptopentanone (**7**), 2,5-dimethyl-4-hydroxy-2,3-dihydrothiophen-3-one (**39**), 2-methyl-1,3-dithiolane (**43**), 3,5-dimethyl-1,2,4-trithiolane (**46**), 3-methyl-1,2,4-trithiane (**48**) and thiazole (**52**).

Thermal degradation of thiamin produces several important compounds.[48,107–109] van der Linde *et al.* reported five primary products, the main component being 4-methyl-5-(2-hydroxyethyl)-thiazole, which is responsible for the formation of several thiazoles on further degradation.[48] The other sulphur-containing primary product is 5-hydroxy-3-mercaptopentanone, a key intermediate compound giving rise to a number of aliphatic sulphur compounds, furans and thiophens.[48] Two of these were also present in the previously discussed reaction products of HMFone/H_2S,[49,86,87] namely 2-methyl-tetrahydrothiophen-3-one and the meaty 2-methyl-4,5-dihydrofuran-3-thiol (**19**). Based on the postulated mechanism for the HMFone/H_2S reaction,[49] it is surprising that another meaty thermal degradation product of thiamin was not identified in the HMFone/H_2S reaction,[49,86,87] namely 2-methylfuran-3-thiol (**12**).

Hartman *et al.* recently degraded thiamin at 135°C/30 min in water (pH 2·3) and also in propane-1,2-diol.[107] Few decomposition products were formed from the diol system but several carbonyls, furanoids, thiophenoids, thiazoles and aliphatic sulphur compounds were isolated from the aqueous reaction. In addition a novel compound was reported, i.e. 3-methyl-4-oxo-1,2-dithiane, also present (and in enhanced concentration) from a model system of cystine/ascorbic acid/thiamin.[107] Reineccius and Liardon have also recently studied the volatile products from thiamin but at lower temperatures (40°, 60°, 90°) and at pH 5, 7 and 9 respectively.[108] At pH 5 and 7, 2-methylfuran-3-thiol (**12**), bis-(2-methyl-3-furyl) disulphide (**24**) and various thiophens were the major products, but at pH 9 the meaty compounds **12** and **24** were not significant and the thiophens predominated.

Some bicyclo compounds (**A**, **B**, **C**) were identified in these two recent publications,[107,108] compounds **B** and **C** by Hartman *et al.*[107] and compounds **A** and **B** by Reineccius and Liardon, and formed only at the higher temperature used.[108] These three

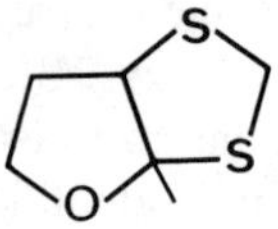

A

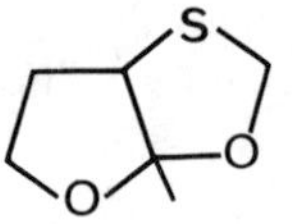

B

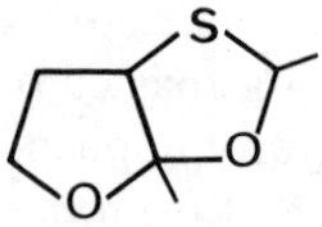

C

compounds have been reported previously as photodegradation products of thiamin and implicated in the rather undesirable, characteristic odour of thiamin itself.[110–113] Compounds **B** and **C** were identified by van Dort *et al.* who described their odours as sulphurous, metallic, green and sweet;[110] their odour thresholds in water are 1 part per 10^7 parts and 1 part per 10^9 parts respectively. They did not identify compound **A** (previously reported by Seifert *et al.*[111]) as a photodegradation product but they synthesised it, and it possessed an odour similar to that of **B** and **C**. Buttery *et al.* later reported that their GC peak, originally characterised as **A**, was in fact a multicomponent peak and, although **A** was a major component, the 'peak odour' was shown to be due to bis-(2-methyl-3-furyl) disulphide (**24**).[113]

Figure 2 compounds reported from thermally degraded thiamin include 3-mercaptopentanone (**7**),[48,107] 2-methylfuran-3-thiol

(**12**),[48,107,108] 2-methyl-4,5-dihydrofuran-3-thiol (**19**),[48,107] 2-methyl-tetrahydrofuran-3-thiol (**21**),[107] bis-(2-methyl-3-furyl) disulphide (**24**),[107,108] 4,5-dimethylthiazole (**54**)[48,107] and 2,5-dimethylfuran-3-thiol (**14**).[107] It is difficult to rationalise the formation of this 2,5-dimethyl compound (**14**) from the key intermediate 5-hydroxy-3-mercaptopentanone (see Fig. 7). If it does form, then potentially, one also expects bis-(2,5-dimethyl-3-furyl) disulphide (**28**), i.e. the dimer of **14** and also bis-(2,5-dimethyl-3-furyl) sulphide (**27**). Although no published reports of the reaction of thiamin/methionine could be traced, the probability is that compound **12**, in the presence of methanethiol, would generate the meaty compounds 2-methyl-3-(methylthio)furan (**15**), 2-methyl-3-(methyldithio)furan (**16**), bis-(2-methyl-3-furyl) trisulphide (**25**) and bis-(2-methyl-3-furyl) tetrasulphide (**26**). The series of reactions just described are summarised in Fig. 8.

This family of furans with sulphur substituents in the 3-position and typified by compounds **12**, **14–17**, **24–28**, are likely to be of extreme importance in cooked meat aromas. Many have been synthesised and patented, and have been described sensorially by Evers *et al.*[44] and by Tressl and Silwar.[50] The odour threshold of bis-(2-methyl-3-furyl) disulphide (**24**) was recently reported as 2

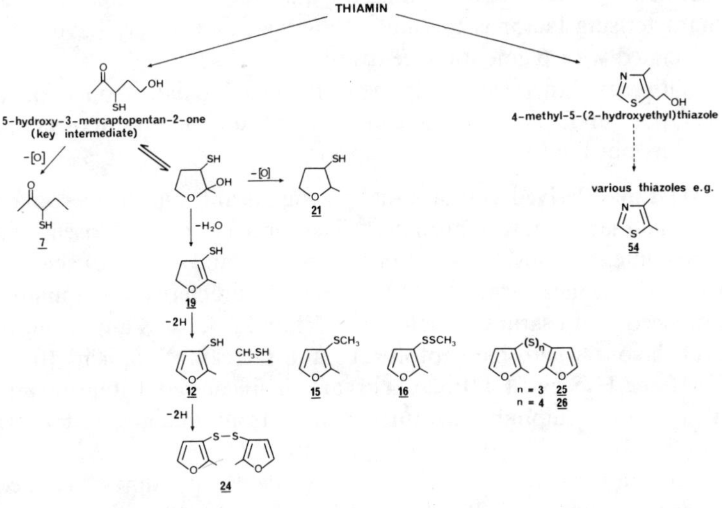

FIG. 8. Schematic representation of the formation of some of the meaty compounds of Fig. 2 from thiamin.

parts per 10^{14} parts water.[113] This is one of the lowest odour thresholds known and is about one-hundred times lower than that of 2-(2-methylpropyl)-3-methoxypyrazine, the character impact compound of bell peppers.[113] Similarly, low odour thresholds in water are reported for 2-methyl-3-(methylthio)furan (**15**) and for 2-methyl-3-(methyldithio)furan (**16**), these being 0·05 and 0·01 ppb respectively.[50] Related structures are likely to be potent odorants too. It follows therefore that only minute traces of these types of compounds need be present for them to be aroma effective, creating enormous analytical difficulties for their detection. One of these compounds, i.e. 2-methyl-3-(methylthio)furan (**15**), was identified recently in cooked beef aroma and suggested as a meaty character impact compound.[40] The same compound (**15**), together with 2-methylfuran-3-thiol (**12**), 2-methyl-3-(methyldithio)furan (**16**) and a bis-(2-methylfuryl) disulphide, were also identified recently in the volatiles from a heated yeast extract composition,[114] the empirical use of which for simulated meat flavourings is established.[6]

Cooked Beef vs. Pork vs. Lamb/Mutton Aromas

In a fascinating recent publication, Puckey and Jones successfully discriminated various meat species using direct probe mass spectrometry of powdered freeze-dried aqueous homogenates of muscle, followed by multivariate statistical analysis of the data.[115] The characterising factor is unknown, however, and it may have been associated with pigment concentration.

With regard to aroma components identified to date, from cooked beef, pork and lamb/mutton, a clear species differentiation does not emerge, but the following facts apply:

(1) Lipid-derived volatiles make a significant impact on cooked pork and lamb/mutton aromas.[34] The former is characterised by a high content of unsaturated (-enoic, -dienoic and -trienoic) acids,[34] whilst sheep meat aromas contain highly characteristic and unique branched chain saturated acids.[8,32,34] Lipid is far less important in beef flavour where mercaptofurans and mercaptothiophens (from HMFone/H_2S and HDFone/H_2S) are significant contributors, and also various sulphur-substituted furans from thiamin and other reactions.

(2) Lamb/mutton aromas contain many alkylpyridines[34] formed from the reaction of alka-2,4-dienals with NH_3.[116] Several appear to be species-specific and a feasible explanation of this was recently

proposed by Baines and Mlotkiewicz in that lamb has a higher concentration of amino acids and a lower content of sugars than either beef or pork.[9] Thus, sugar concentration becomes a limiting factor as far as Maillard reactions are concerned, and when sugars and diketones are depleted, the excess amino acids remaining are thermally degraded forming NH_3.

(3) Lamb/mutton aromas have a high concentration of 3,5-dimethyl-1,2,4-trithiolane (**46**) and thialdine (**68**)[117] and other sulphur compounds too, e.g. 3,6-dimethyl-1,2,4,5-tetrathiane, ethane-1,1-dithiol, bis-(1-mercaptoethyl) sulphide, bis(mercaptomethyl) sulphide and 1,2,3,5,6-pentathiepane, i.e. lenthionine.[34] So far, the last four compounds are specific to sheep meat aromas, and could be simple reaction products from acetaldehyde or formaldehyde with H_2S.[10] Sheep meat does contain a higher concentration of sulphur amino acids than beef or pork, and it also liberates H_2S more rapidly on heating.[8]

(4) This combined presence of rather undesirable odours from branched chain saturated acids, alkylpyridines and high levels of sulphur compounds has been suggested to explain the rejection of lamb/mutton by some consumers.[8,32,116,117]

(5) A characteristic pork rind/fried bacon odour is reportedly due to 3,5-di-(2-methylpropyl)-1,2,4-trithiolane (**47**) and 2,4,6-tri-(2-methylpropyl)tetrahydro-1,3,5-dithiazine (**71**)[64] as mentioned earlier. However, the fact that pork contains a higher concentration of thiamin than beef or lamb/mutton has not been reflected yet in the volatiles identified.

SIMULATED MEAT FLAVOURINGS

The technological basis of meat flavours evolves from the precursors and reactions already described. Patents published since 1960 indicate three basic classes of simulated meat flavourings, although a combined use of two or more of the classes is common:[6] (a) 'reaction products' formed by heating relevant precursors; (b) natural meat-based flavourings; and (c) meat aroma chemicals used singly or in admixtures.

'Reaction Product' Meat Flavourings

Ingredients selected for thermal processing often consist of an amino acid and/or sulphur source combined with a reducing sugar

TABLE 4

Some Ingredients Patented for 'Reaction Product' Meat Flavourings[a]

1. *Amino acid/amine sources*
   cysteine, cystine, methionine, alanine, glycine, lysine, arginine, histidine, tryptophan, proline, valine, glutamic acid, glutamine, aspartic acid, glutathione, other S-containing peptides, HVP (groundnut, soybean, wheat/maize gluten), other hydrolysed proteins (casein, egg, fish, blood, liver, bone,[118] collagen[119]), yeast extract, autolysed yeast, meat extract, taurine, pyrrolidone carboxylic acid
2. *Sulphur sources*
   H_2S, cysteine, cystine, methionine, glutathione, thiamin, inorganic sulphides, organic thiols and sulphides, 2-mercaptoethanol derivatives e.g. mercaptoacetaldehyde and/or its dimer 2,5-dihydroxy-1,4-dithiane, 5-hydroxy-3-mercaptopentanone, 3-mercaptopropan-1-ol, 4,5-substituted thiazoles, thiocarbonates, thioamides, 2-mercaptoalkanoic acids/amides, mercaptoalkylamines, aminosulphides, *S*-acetylmercaptosuccinic acid, vegetable extracts, fermented vegetable juice,[120] yeast extract, autolysed yeast, egg protein, meat extract
3. *Reducing sugar (and derivative) sources*
   ribose, xylose, arabinose, glucose, fructose, lactose, sucrose, ribose-5-phosphate, yeast extract, autolysed yeast, HVP, nucleotides, dextrins,[121] pectins, alginates
4. *α-Diketone and potential α-diketone sources*
   butanedione, pentane-2,3-dione, pyruvaldehyde, pyruvic acid, glyceraldehyde, glyoxal, dihydroxyacetone, α-ketobutyric acid, heptane-3,4-dione-2,5-diacetate, HMFone, HDFone and related derivatives, ascorbic acid, 5-ketogluconic acid,[94] cyclotene, maltol, lactic acid, glycollic acid, malic acid, tartaric acid, protein hydrolysates
5. *Aldehyde sources*
   acetaldehyde, propanal, butanal, methylpropanal, C_5 to C_8 alkanals, HVP
6. *Flavour enhancer sources*
   MSG, IMP, GMP, yeast extract, autolysed yeast, HVP, 2-furfurylthioinosine-5′-phosphate, 2-allyloxyinosine-5′-phosphate, 2-(lower alkoxy)inosine-5′-phosphate, 2-benzylthioinosine-5′-phosphate, 4-glucosylgluconic acid i.e. maltobionic acid, cyclotene
7. *pH regulators*
   inorganic acids (e.g. HCl, H_2SO_4, H_3PO_4), organic acids (e.g. succinic, citric, lactic, malic, tartaric, acetic, propanoic), amino acids (e.g. valine, glycine, glutamic acid)
8. *Fat*
   beef, chicken, coconut, triglycerides, fatty acids or esters
9. *Inorganic salts*
   chlorides, phosphates
10. *Spices*

[a] Where no supporting reference is quoted, the data derive from the summarised patent literature for 1960–80;[6] some additional data since 1980 are referenced.[94,118–121]

or carbonyl compound(s). As already shown, however, alternatives are possible and Table 4 summarises the main ingredients patented in 'reaction product' meat flavourings.[6,94,118–121] Hshieh *et al.* have reported their formulation of a model meat flavour system using sensory analysis in conjunction with the surface response method for testing the contribution and optimum concentrations of ingredients.[4] The final model (% dry weight in brackets) consisted of: glucose (6·4), ribose (4·3), xylose (6·4), alanine (2·5), MSG (2·1), cysteine (2·1), taurine (2·4), serine (2·2), leucine (1·0), isoleucine (1·0), methionine (1·2), glycine (0·8), arginine (0·6), IMP (2·7), gelatin (53·5) and NaCl (10·7). Gelatin masked a strong sulphurous aroma, increased brothiness and improved mouthfeel. Fat (69·7) improved the aroma but was quoted as optional because of rancidity problems. The model was equal or superior to five commercial meat flavourings tested, and nearly equal to an authentic extract prepared from beef. Whilst the most significant amino acids in this model were cysteine, alanine, glutamic acid, serine and taurine, Schrödter *et al.*, in optimising the levels of nine amino acids heated with glucose to give roast beef aroma, stressed the importance of cysteine, glutamic acid, aspartic acid, proline and arginine.[122]

Natural Meat-based Flavourings

These are products in which natural meat precursors make a significant contribution to the flavour, although other ingredients may be added. Meat slurries, extracts, dialysates, liquified meat or enzyme digests of meat have been used.[6,119,123] Yeast hydrolysates are often present and autolysis results in hydrolysis of proteins and nucleotides, together with the conversion of hexoses and pentoses into their phosphates.[6]

Meat Aroma Chemicals

Many of these are presented in Fig. 2 and discussed already. A few selected additional examples are as follows:

3-Mercapto-2-methylpropan-1-ol[124]
Alkanedithiols[6]
Mercaptosulphides[6]
Mercaptoketones[6]
Cycloalkylthiols[6]
3-Furylthioesters[6]

Bis(tetrahydrofuryl-2) sulphide and derivatives[6]
Bis(thienyl) sulphide derivatives[6]
4,5-Dimethyl-2-(2-methylthioethyl)-1,3-oxathiolane[125]
Methyl(methylthioalkyl)-1,3-dithiolanes[126]
4 or 5-Methyl-2-(2,6-dimethyl-1,5-heptadienyl)-1,3-dithiolane[127]
3,5-Dispiro-1,2,4-trithiolanes[128]
Acylpyrimidines[129]

Very fine concentration tuning is called for in most cases, since the vast majority of these compounds are highly potent odorants conferring desirable meatiness only at certain restricted dilutions.

CONCLUSION

The vast majority of compounds described as meaty are sulphur-containing and of low odour thresholds. The reaction products of HMFone/H_2S are of undoubted importance, together with certain furans substituted with sulphur in the 3-position, e.g. **12**, **15**, **16**, **24–26**. Their high aroma potency and consequent presence in such trace amounts explains why analysts have, so far, failed to detect the majority of them in natural cooked meat aromas. Others await characterisation, both from natural meat and from relevant model systems based on naturally occurring precursors. Consumers are likely to steer future technology even more towards the use of components of natural origin, and gaps in our current knowledge need to be filled. As stated by Ohloff *et al.*,[55] we also know very little about the stimulation of our chemical senses and about the physicochemical basis of synergism. We therefore need more information to define scientifically the most important flavours and how they work, and technically to create better flavours to please a constantly growing and more discriminating population.

REFERENCES

1. Mottram, D. S. and Edwards, R. A. (1983). *J. Sci. Food Agric.*, **34**, 517–22.
2. Pearson, A. M. and Gray, J. I. (1983). In: *The Maillard Reaction in Foods and Nutrition*, Eds Waller, G. R. and Feather, M. S., ACS Symp. Ser. 215, Amer. Chem. Soc., Washington, DC, 287–300.

3. Alabran, D. M. (1982). *J. Agric. Food Chem.*, **30**, 486–8.
4. Hsieh, Y. P. C., Pearson, A. M. and Magee, W. T. (1980). *J. Food Sci.*, **45**, 1125–30, 1135.
5. Katz, I. (1981). In: *Flavour Research: Recent Advances*, Eds Teranishi, R., Flath, R. A. and Sugisawa, H., Marcel Dekker, New York, 217–29.
6. MacLeod, G. and Seyyedain-Ardebili, M. (1981). *CRC Crit. Rev. Food Sci. Nutr.*, **14**, 309–437.
7. Bailey, M. E. (1983). In: *The Maillard Reaction in Foods and Nutrition*, Eds Waller, G. R. and Feather, M. S., ACS Symp. Ser. 215, Amer. Chem. Soc., Washington, DC, 169–84.
8. IFT (1983). Symposium, 42nd Ann. Meeting, *Food Technol.*, **37**, 225–68.
9. Baines, D. A. and Mlotkiewicz, J. A. (1984). In: *Recent Advances in the Chemistry of Meat*, Ed. Bailey, A. J., Roy. Soc. Chem., London, 119–64.
10. MacLeod, G. (1984). *Proc. Inst. Food Sci. Technol.*, **17**, 184–97.
11. Ney, K. H. (1979). In: *Food Taste Chemistry*, Ed. Boudreau, J. C., ACS Symp. Ser. 115, Amer. Chem. Soc., Washington, DC, 149–73.
12. Motono, M. (1982). In: *Chemistry of Foods and Beverages: Recent Developments*, Eds Charalambous, G. and Inglett, G., Academic Press, New York, 181–94.
13. Yamaguchi, S. (1979). In: *Food Taste Chemistry*, Ed. Boudreau, J. C., ACS Symp. Ser. 115, Amer. Chem. Soc., Washington, DC, 33–51.
14. Mabrouk, A. F. (1976). In: *Phenolic, Sulphur and Nitrogen Compounds in Food Flavours*, Eds Charalambous, G. and Katz, I., ACS Symp. Ser. 26, Amer. Chem. Soc., Washington, DC, 146–83.
15. Schutte, L. (1976). In: *Phenolic, Sulphur and Nitrogen Compounds in Food Flavours*, Eds Charalambous, G. and Katz, I., ACS Symp. Ser. 26, Amer. Chem. Soc., Washington, DC, 96–113.
16. Kuninaka, A. (1981). In: *Flavour Research: Recent Advances*, Eds Teranishi, R., Flath, R. A. and Sugisawa, H., Marcel Dekker, New York, 305–53.
17. Solms, J. (1968). *Fleischwirtschaft*, **48**, 287–91.
18. Snider, S. and Baldwin, R. E. (1981). *J. Food Sci.*, **46**, 1801–4.
19. Kirova, E. and Peschevska, M. (1982). *Nährung*, **26**, 669–74.
20. Nguyen, T. T. and Sporns, P. (1985). *J. Food Sci.*, **50**, 812–4, 822.
21. Gayte-Sorbier, A., Airaudo, C. B. and Armand, P. (1985). *J. Food Sci.*, **50**, 350–2, 360.
22. Macy, R. L. Jr, Naumann, H. D. and Bailey, M. E. (1970). *J. Food Sci.*, **35**, 83–7.
23. Pearson, A. M., Wolzak, A. M. and Gray, J. I. (1983). *J. Food Biochem.*, **7**, 189–210.
24. Macy, R. L. Jr, Naumann, H. D. and Bailey, M. E. (1964). *J. Food Sci.*, **29**, 142–8.
25. May, C. G. (1974). *Food Trade Rev.*, **44**, 7–14.
26. Morton, I. D., Akroyd, P. and May, C. G. (1960). Brit. Pat. 836,694.

27. Love, J. D. (1983). *Food Technol.*, **37**, July, 117–20, 129.
28. Mottram, D. S., Edwards, R. A. and Macfie, H. J. H. (1982). *J. Sci. Food Agric.*, **33**, 934–44.
29. Winstanley, M. (1985). *Food*, **7**(3), 34–5.
30. Mottram, D. S. (1977–9). *ARC/MRI (Bristol), Biennial Rep.*, 87–8.
31. Brennand, C. P. and Lindsay, R. C. (1982). *Lebensm. Wiss. Technol.*, **15**, 249–52.
32. Wong, E., Nixon, L. N. and Johnson, C. B. (1975). *J. Agric. Food Chem.*, **23**, 495–8.
33. Lingnert, H. and Eriksson, C. E. (1981). In: *Progress in Food and Nutrition Science*. Vol. 5. *Maillard Reactions in Food*, Ed. Eriksson, C., Pergamon Press, Oxford, 453–66.
34. van Straten, S. and Maarse, H. (1983). *Volatile Compounds in Foods*, 5th edn, Division for Nutrition and Food Research TNO, The Netherlands.
35. Ching, J. C.-Y. (1979). *Volatile Flavour Compounds from Beef and Beef Constituents*, University of Missouri.
36. Galt, A. M. and MacLeod, G. (1984). *J. Agric. Food Chem.*, **32**, 59–64.
37. Yamaguchi, K., Shudo, K., Okamoto, T., Sugimura, T. and Kosuge, T. (1980). *Gann*, **71**, 745–6.
38. Wasserman, A. E. (1979). *J. Food Sci.*, **44**, 6–11.
39. Hartman, G. J., Jin, Q. Z., Collins, G. J., Lee, K. N., Ho, C.-T. and Chang, S. S. (1983). *J. Agric. Food Chem.*, **31**, 1030–3.
40. MacLeod, G. and Ames, J. M. (1986). *Chem. and Ind.*, 175–7.
41. Brinkman, H., Copier, H., de Leuw, J. M. and Sing, B. T. (1972). *J. Agric. Food Chem.*, **20**, 177–81.
42. Boelens, M., van der Linde, L., de Valois, P., van Dort, H. and Takken, H. (1974). *J. Agric. Food Chem.*, **22**, 1071–6.
43. Arctander, S. (1969). *Perfume and Flavour Chemicals*, published by the author, New Jersey.
44. Evers, W. J., Heinsohn, H. H. Jr, Mayers, B. J. and Sanderson, A. (1976). In: *Phenolic, Sulphur and Nitrogen Compounds in Food Flavours*, Eds Charalambous, G. and Katz, I., ACS Symp. Ser. 26, Amer. Chem. Soc., Washington, DC, 184–93.
45. Nishimura, O., Mihara, S. and Shibamoto, T. (1980). *J. Agric. Food Chem.*, **28**, 39–43.
46. Maga, J. A. (1976). *CRC Crit. Rev. Food Sci. Nutr.*, **7**, 147–92.
47. Ohloff, G. and Flament, I. (1978). *Heterocycles*, **2**, 663–95.
48. van der Linde, L. M., van Dort, J. M., de Valois, P., Boelens, H. and de Rijke, D. (1979). In: *Progress in Flavour Research*, Eds Land, D. G. and Nursten, H. E., Applied Science, London, 219–24.
49. van den Ouweland, G. A. M. and Peer, H. G. (1975). *J. Agric. Food Chem.*, **23**, 501–5.
50. Tressl, R. and Silwar, R. (1981). *J. Agric. Food Chem.*, **29**, 1078–82.
51. IFF Inc. (1979). Brit. Pat. 1,543,653.
52. Rödel, W. and Kruse, H. P. (1980). *Nährung*, **24**, 129–39.

53. Shibamoto, T. (1980). *J. Agric. Food Chem.*, **28**, 237–43.
54. van der Heijden, A. and Schutte, L. (1974). US Pat. 3,787,473.
55. Ohloff, G., Flament, I. and Pickenhagen, W. (1985). *Food Rev. Int.*, **1**, 99–148.
56. King, B., Demole, E. and Thomas, A. F. (1972). DOS 2,214,540.
57. Shu, C.-K., Hagedorn, M. L., Mookherjee, B. D. and Ho, C.-T. (1985). *J. Agric. Food Chem.*, **33**, 638–41.
58. Tressl, R., Helak, B., Grünewald, H.-G. and Silwar, R. (1983). In: *Colloque International sur les Arômes Alimentaires*, Eds Adda, J. and Richard, H., Technique et Documentation-Lavoisier, Paris, 207–30.
59. Shu, C.-K., Mookherjee, B. D. and Vock, M. (1981). US Pat. 4,259,507.
60. Wilson, R. A., Mussinan, C. J., Katz, I., Giacino, C., Sanderson, A. and Shuster, E. J. (1975). US Pat. 3,863,013.
61. Sakaguchi, M. and Shibamoto, T. (1978). *J. Agric. Food Chem.*, **26**, 1260–2.
62. Kubota, K., Kobayashi, A. and Yamanishi, T. (1980). *Agric. Biol. Chem.*, **44**, 2677–82.
63. Self, R., Casey, J. C. and Swain, T. (1963). *Chem. and Ind.*, 863–4.
64. Shu, C.-K., Mookherjee, B. D., Bondarovich, H. A. and Hagedorn, M. L. (1985). *J. Agric. Food Chem.*, **33**, 130–2.
65. Wilson, R. A., Vock, M. H., Katz, I. and Shuster, E. J. (1974). Brit. Pat. 1,364,747.
66. Shu, C.-K., Hagedorn, M. L., Mookherjee, B. D. and Ho, C.-T. (1985). *J. Agric. Food Chem.*, **33**, 442–6.
67. Baltes, W. (1979). *Deutsch. Lebensm. Rundsch.*, **75**, 2–7.
68. Furia, T. W. and Bellanca, N. (Eds) (1975). *Fenaroli's Handbook of Flavour Ingredients*, 2nd edn, Vol. 1, CRC Press, Ohio.
69. Vernin, G. (1979). *Parfum Cosmétiq. Arômes*, **29**, June/July, 77–87.
70. Pittet, A. and Hruza, D. (1974). *J. Agric. Food Chem.*, **22**, 264–9.
71. Mussinan, C. J., Wilson, R. A., Katz, I., Hruza, A. and Vock, M. H. (1976). In: *Phenolic, Sulphur and Nitrogen Compounds in Food Flavours*, Eds Charalambous, G. and Katz, I., ACS Symp. Ser. 26, Amer. Chem. Soc., Washington, DC, 133–45.
72. Vernin, G. (1982). *The Chemistry of Heterocyclic Flavouring and Aroma Compounds*, Ellis Horwood, Chichester.
73. Vock, M., Giacino, C., Hruza, A., Withycombe, D., Mookherjee, B. D. and Mussinan, C. (1981). US Pat. 4,243,688.
74. Calabretta, P. (1978). *Perfum. Flavor.*, **3**, June/July, 33–42.
75. Tressl, R., Renner, R., Kossa, T. and Köppler, H. (1977). *Europ. Brew. Conv. Proc. 16th Congr.*, 693–707.
76. Mecchi, E. P., Pippen, E. L. and Lineweaver, H. (1964). *J. Food Sci.*, **29**, 393–9.
77. Kato, S., Kurata, T. and Fujimaki, M. (1973). *Agric. Biol. Chem.*, **37**, 539–44.
78. Ho, C.-T. and Hartman, G. J. (1982). *J. Agric. Food Chem.*, **30**, 793–4.

79. DE RIJKE, D., VAN DORT, J. M. and BOELENS, H. (1981). In: *Flavour '81*, 3rd Weurman Symposium, Ed. Schreier, P., de Gruyter, Berlin, 417–31.
80. SCANLAN, R. A., KAYSER, S. G., LIBBEY, L. M. and MORGAN, M. E. (1973). *J. Agric. Food Chem.*, **21**, 673–5.
81. MULDERS, E. J. (1973). *Z. Lebensm. Unters. Forsch.*, **152**, 193–201.
82. TRESSL, R., BAHRI, D., HOLZER, M. and KOSSA, T. (1977). *J. Agric. Food Chem.*, **25**, 459–63.
83. SHU, C.-K., HAGEDORN, M. L., MOOKHERJEE, B. D. and HO, C.-T. (1985). *J. Agric. Food Chem.*, **33**, 438–42.
84. TAKKEN, H. J., VAN DER LINDE, L. M., DE VALOIS, P. J., VAN DORT, H. M. and BOELENS, M. (1976). In: *Phenolic, Sulphur and Nitrogen Compounds in Food Flavours*, Eds Charalambous, G. and Katz, I., ACS Symp. Ser. 26, Amer. Chem. Soc., Washington, DC, 114–21.
85. KAWAI, T., MAMORU, I. and SAKAGUCHI, M. (1985). *J. Agric. Food Chem.*, **33**, 393–7.
86. VAN DEN OUWELAND, G. A. M. and PEER, H. G. (1979). US Pat. 4,134,901.
87. VAN DEN OUWELAND, G. A. M., OLSMAN, H. and PEER, H. G. (1978). In: *Agricultural and Food Chemistry: Past, Present and Future*, Ed. Teranishi, R., AVI, Westport, 292–314.
88. TONSBEEK, C. H. T., PLANCKEN, A. J. and WEERDHOF, T. V. D. (1968). *J. Agric. Food Chem.*, **16**, 1016–21.
89. TONSBEEK, C. H. T., KOENDERS, E. B., VAN DER ZIJDEN, A. S. M. and LOSEKOOT, J. A. (1969). *J. Agric. Food Chem.*, **17**, 397–400.
90. PEER, H. G. and VAN DEN OUWELAND, G. A. M. (1968). *Rec. Trav. Chim.*, **87**, 1017–20.
91. HICKS, K. B. and FEATHER, M. S. (1975). *J. Agric. Food Chem.*, **23**, 957–60.
92. HICKS, K. B., HARRIS, D. W., FEATHER, M. S. and LOEPPKY, R. N. (1974). *J. Agric. Food Chem.*, **22**, 724–5.
93. PEER, H. G., VAN DEN OUWELAND, G. A. M. and DE GROOT, C. N. (1968). *Rec. Trav. Chim.*, **87**, 1011–6.
94. DE ROOIJ, J. F. M. (1984). US Pat. 4,464,409.
95. MACY, R. L. JR, NAUMANN, H. D. and BAILEY, M. E. (1970). *J. Food Sci.*, **35**, 81–3.
96. JONES, N. R. (1969). *J. Agric. Food Chem.*, **17**, 712–16.
97. KHYM, J. X., DOHERTY, D. G. and COHN, W. E. (1954). *J. Amer. Chem. Soc.*, **76**, 5523–30.
98. TRESSL, R., GRÜNEWALD, K. G., SILWAR, R. and BAHRI, D. (1979). In: *Progress in Flavour Research*, Eds Land, D. G. and Nursten, H. E., Applied Science, London, 197–213.
99. HEYNS, K., STUTE, R. and PAULSEN, H. (1966). *Carbohydr. Res.*, **2**, 132–49.
100. PISAMITSKI, A. F., EGOROV, I. A., EGOFAROVA, R. K. and ERYGIN, G. D. (1979). *Appl. Biochem. Microbiol.*, **15**, 225–31.
101. VAN DEN OUWELAND, G. A. M. and SWAINE, R. L. JR (1980). *Perfum. Flavor.*, **5**, 15–20.
102. TAKEI, Y. (1977). *Agric. Biol. Chem.*, **41**, 2361–8.

103. Doornbos, T., van den Ouweland, G. A. M. and Tjan, S. B. (1981). In: *Progress in Food and Nutrition Science. Vol. 5. Maillard Reactions in Food*, Ed. Eriksson, C., Pergamon Press, Oxford, 57–63.
104. Hirvi, T., Honkanen, E. and Pyysalo, T. (1980). *Lebensm. Wiss. Technol.*, **13**, 324–5.
105. Shu, C.-K., Mookherjee, B. D. and Ho, C.-T. (1985). *J. Agric. Food Chem.*, **33**, 446–8.
106. Bodrero, K. O., Pearson, A. M. and Magee, W. T. (1981). *J. Food Sci.*, **46**, 26–31.
107. Hartman, G. J., Carlin, J. T., Scheide, J. D. and Ho, C.-T. (1984). *J. Agric. Food Chem.*, **32**, 1015–18.
108. Reineccius, G. A. and Liardon, R. (1985). In: *Topics in Flavour Research*, Eds Berger, R. G., Nitz, S. and Schreier, P., Eichhorn-Hangenham, 125–36.
109. Dwivedi, B. K. and Arnold, R. G. (1973). *J. Agric. Food Chem.*, **21**, 54–60.
110. van Dort, H. M., van der Linde, L. M. and de Rijke, D. (1984). *J. Agric. Food Chem.*, **32**, 454–7.
111. Seifert, R. M., Buttery, R. G., Lundin, R. E., Haddon, W. F. and Benson, M. (1978). *J. Agric. Food Chem.*, **26**, 1173–76.
112. Buttery, R. G., Seifert, R. M., Turnbaugh, J. G., Guadagni, D. G. and Ling, L. C. (1981). *J. Agric. Food Chem.*, **29**, 183–5.
113. Buttery, R. G., Haddon, W. F., Seifert, R. M. and Turnbaugh, J. G. (1984). *J. Agric. Food Chem.*, **32**, 674–6.
114. Ames, J. M. and MacLeod, G. (1985). *J. Food Sci.*, **50**, 125–31, 135.
115. Puckey, D. J. and Jones, S. J. (1984). *J. Analyt. Appl. Pyrol.*, **6**, 157–69.
116. Buttery, R. G., Ling, L. C., Teranishi, R. and Mon, T. R. (1977). *J. Agric. Food Chem.*, **25**, 1227–9.
117. Nixon, L. N., Wong, E., Johnson, C. B. and Birch, E. J. (1979). *J. Agric. Food Chem.*, **27**, 355–9.
118. Behnke, U., Ackermann, E., Kroll, J., Weibelzahl, H. and Kalinna, B. (1985). GDR Pat. 217,980.
119. Kyowa Hakko Kogyo (1981). Jap. Pat. 5,646,813.
120. Hsu, J. Y. (1984). US Pat. 4,464,407.
121. Lieske, B., Konrad, G. and Schulze, W. (1985). GDR Pat. 217,981.
122. Schrödter, R., Wölm, G. and Schliemann, J. (1982). *Nährung*, **26**, 625–31.
123. Huth, H. and Schum, H. (1983). Ger. Pat. Appl. 32,06,587.
124. Mikhailova, T. V., Gren, A. I., Vysotskaja, L. E., Misharina, T. A., Vitt, S. V. and Golovnya, R. V. (1985). *Nährung*, **29**, 671–80.
125. Pittet, A. O., Courtney, T. F. Jr and Muralidhara, R. (1985). US Pat. 4,515,966.
126. Pittet, A. O., Courtney, T. F. Jr and Muralidhara, R. (1985). US Pat. 4,515,967.
127. Pittet, A. O., Vock, M. H., Courtney, T. F. Jr and Muralidhara, R. (1984). US Pat. 4,464,408.
128. van den Bosch (1985). Ger. Pat. 26,30,327.
129. Firmenich SA (1980). Brit. Pat. 1,561,121.

13

The Influence of Legislation on Development in the Flavour Industry

JACK KNIGHTS

PFW Ltd, Greenford, Middx, UK

ABSTRACT

The development of the flavour industry since the discovery of gas–liquid chromatography some thirty years ago has been rapid and to the benefit of society at large. The development has been influenced by the differing systems of legislation to a lesser or greater extent. Before it is possible to examine the scale of the problem of legislation it is necessary to define flavourings and ingredients and to group them appropriately. It is demonstrated that, even in the UK which boasts no flavour legislation, 95% by weight of the ingredients are regulated by positive lists. The remaining 5% consists of more than 3000 materials of flavour potential divided into naturals, nature identicals and artificials. Each of these groups is discussed and the rationale behind nature identity explained by the introduction of the concept of food predominance. The number of materials in each group is estimated and quantitative data on their average consumption are presented. The analytical difficulties in differentiating between natural and nature identical ingredients are examined. The existence of non-defined flavourings, mainly process and microbiological flavours, is examined from the viewpoint of the analytical problem they pose to the legislator. Specific flavour legislation can be either positive or negative. The merits of each system are set out and a mixed system is proposed as a workable compromise. The impact of positive legislation is illustrated with reference to the USA, and the consequences of its implementation on flavour research and development are discussed in some detail. Particular reference is made to the development of natural undefined flavourings. It is not possible to find an example of total negative legislation, so the mixed system implemented in West Germany is discussed. This system is less constraining to research and development and more advantageous to the final consumer. In the UK the burden of safety rests with the flavour manufacturer and it is concluded that this is the most acceptable system. As a fall-back the IOFI mixed system would be acceptable and not present any hazard to the consumer.

INTRODUCTION

There is probably no industry that has as great a reputation for secrecy as the flavour industry. This view is justified to a considerable extent and comes about because the products are generally simple mixtures of ingredients, the know-how being contained in the quantitative and qualitative make-up of these mixtures. For this reason every effort has been made to keep the identity of the characteristic ingredients secret and their relationship to the final flavouring confidential.

Thirty years ago, with the introduction of gas–liquid chromatography (GLC), it was forecast that all the secrets of the flavour industry would be laid bare and that no company would be able to introduce a new flavouring withouts its competitors being able to determine the constitution, the business degenerating into a series of price wars.

In the event, the introduction of GLC has resulted in an increase in the knowledge of natural food products and a proliferation of the flavouring ingredients available to the flavourist. Many of these materials are of such flavour potency that they are used at a few parts per million in a finished flavouring (a level at which they are in much smaller proportion than many of the impurities in the major ingredients), making them very difficult to detect, even with modern GC/MS equipment.

The important contribution made by modern GC/MS techniques is the primary vector in the improvement of the character of the flavourings presented to the food manufacturer and thus to the consumer. This has resulted in an improvement of the acceptability of processed food products, so important in an industrialised society.

In this paper it is proposed to look at the way in which differing systems of legislation have affected the efficiency with which this flavour improvement has been effected. This will be examined in three parts.

THE SCALE OF THE LEGISLATIVE PROBLEM FOR FLAVOURINGS

In order to start discussing the scale of the problem it is necessary to have a definition of what is meant by flavourings and I intend to use that of IOFI: Flavouring means a concentrated preparation, with or without flavour adjuncts, used to impart flavour, with the exception of only salty, sweet or acid tastes. It is not intended to be consumed as such.[1]

In practice most flavourings are mixtures of flavouring ingredients, sometimes together with natural food substances, often with a solvent or support, to provide a material which has the correct organoleptic and physical properties and the ideal concentration for the final user.

Reviewing the options available for non-flavouring ingredients, one sees the following groups:

(i) Natural food substances, such as fruit juices, sugars, starches, rusk, hydrolysed protein.
(ii) Solvents; usually regulated by permitted list.
(iii) Carriers: gum arabic, modified starches; usually either permitted list or natural food substance.
(iv) Emulsifiers and stabilisers; permitted list.
(v) Antioxidants: BHA, BHT, gallates; permitted list.
(vi) Sequestrants: citric acid, EDTA, phosphates; permitted list.
(vii) Preservatives: sulphur dioxide, benzoates; permitted list.
(viii) pH regulators: acids and their salts; permitted list.

As can be seen from the above list, these products do not prove a difficulty from a legislative viewpoint since they are mainly positive-listed. However, in terms of quantity they represent about 86% of the total attributed to flavours in the UK. Of the total, 9% consists of process flavours which are described later in this paper.

The remaining 5% (or 50 g/person/year) consists of defined natural, nature identical and artificial flavouring substances. Before we proceed further we should define what is meant by natural, nature identical and artificial flavouring ingredients.[1] Natural ingredients can be divided into three types:

(a) Natural aromatic raw material: vegetable or animal product used for its flavouring properties, either as such or processed and suitable for human consumption in the form in which it is used. Examples are fruits, fruit juices, spices, herbs, cheese and vinegar.
(b) Natural flavour concentrate: concentrated preparations obtained exclusively by physical methods from a natural aromatic raw material. Examples are essential oils, oleoresins, extracts and distillates.
(c) Natural flavouring substance: defined chemical component isolated from a natural aromatic raw material by exclusively physical methods. Examples are citral, geraniol, menthol.

A 'nature identical' flavouring substance is obtained by synthesis or isolated by a chemical process from natural aromatic raw material and is chemically identical to a substance present in natural products, intended for human consumption, either processed or not. An 'artificial' flavouring substance has not yet been identified in a natural product intended for human consumption, either processed or not. These are the internationally agreed definitions derived by IOFI.

The rationale behind dividing synthetic flavouring substances into nature identical and artificial is as follows.

Nature identical materials will present no greater hazard to health than the food supply itself, providing they are present at no greater level than in food. This has led to the introduction of the concept of consumption ratio for flavouring substances.[2]

Consumption ratio (CR) is defined as the ratio between the amount being consumed via the natural food supply compared with that consumed via the addition as a nature identical chemical. This means that if the CR is greater than 1 then the material could be termed food-predominant and regarded as having no greater hazard than food itself. As a consequence of this definition it can be seen that artificial flavouring substances have CR's of zero. Most nature identical flavouring substances have CR's greater than 1, but an obvious exception is vanillin whose CR is probably as low as 0·01.

Maybe it would be appropriate to redefine nature identical as meaning a flavouring substance with a CR of 1 or greater, with the consequence that vanillin would then become artificial.

Of the total defined flavouring substances, about 500 are natural with an average consumption of 45 g/person/year consisting largely of citrus and mint oils and pepper oleoresin. The remaining nature identical and artificial flavouring ingredients are consumed at a total maximum level of about 5 g/person/year and consist of more than 3000 materials. The real number of materials is not known since this is a function of the ability of analysts and flavourists to make use of what is found of natural occurrence in food. Later in this paper I will illustrate the increase in analytical food knowledge with time. However, going back to the most important nature identical materials, there are about 32 compounds used at a level of over 500 kg/year in the UK. This level represents an average annual intake of 10 mg/person. There are about a further 100 materials estimated to be used at over 50 kg/year (1 mg/person). The total of artificials (i.e. non-nature identicals) represents a total intake of 0·35 g/person/year and consists of four materials at over 500 kg/year and six at over 50 kg/year. The four are ethyl vanillin (12 tonnes), ethyl maltol (4·5 tonnes), ethyl methyl phenyl glycidate (630 kg) and butyl butyryl lactate (550 kg).[3]

To put the usage of artificial flavouring substances into a world context, IOFI has undertaken a survey of their incorporation into food by the companies comprising membership of national organisations making up the IOFI, which covers most flavour manufacturers outside Eastern Europe. As a result of this survey the usage of the same four materials, ethyl vanillin (161 tonnes), ethyl maltol (46 tonnes), ethyl methyl phenyl glycidate (13·5 tonnes) and butyl butyryl lactate (19·8 tonnes) exceeds 5 tonnes.[4] It should be noted that the individual components of the latter, viz. lactic acid, butyric acid, butyl alcohol, all occur in foodstuffs and that *in vitro* and *in vivo* hydrolyses suggest that the material is metabolised as its component alcohol and acids. In addition, a further 21 artificial compounds are used at more than 1 tonne.

Since the international flavour industry has accepted that regulation of the artificial flavouring ingredients may be by means of a permitted list, the difficulty of flavour legislation resolves itself to dealing with some 500 natural extracts and essential oils and an indeterminate number of nature identicals. The natural extracts and essential oils tend to be well known and quite well documented although many of the less common ones are not well characterised in terms of their constituents. This causes one of the legislative difficulties, since constituents of spice and food extracts and essential oils are identical to the nature identical chemicals used. It is therefore not simple to determine whether, for instance, the cinnamaldehyde present in a finished flavouring or food originates from a raw spice, extract or essential oil, or whether it is added as a simple fine chemical. Whilst in this case it is of little importance, there are many constituents of natural food products which are insufficiently documented for it to be known whether the chemical species present is derived from food or added, and thus whether it is nature identical or artificial. In addition, if one were to envisage a totally positive list system, it would be impossible to be certain that the presence of a non-listed material was due to its being added as such or as part of a natural product.

So far the discussion has only dealt with defined flavour ingredients. Just as it is possible to make a food which does not have the flavour characteristics of its ingredients, for instance toffee, it is possible to make flavours similarly from suitable precursors. These precursors are often natural food materials or their isolates which are processed to emphasise the production of flavour character and represent roughly 9% of the total flavour market. Examples of these kinds of flavours are:

(a) Process flavours, e.g. meat, chocolate, caramel, potato.
(b) Enzymic and microbiological flavours, e.g. cheese, butter, bread.
(c) Pyrolysis flavours, e.g. smoke, grill.

In addition to these mainly modified natural food products there are also chemical reaction products, examples of which are provided by Schiff bases from aldehydes and methyl anthranilate, esterified or reduced essential oils and extracts, e.g. acetylated peppermint oil, ethyl esters of herring oil, hydrogenated spearmint oil. All of the above groups of reaction products result in complex mixtures, most of which have never been documented outside the flavour industry.

In the foregoing part of the paper I have indicated the immense complexity of the materials used by the flavour industry. It is obvious that there is a high degree of uncertainty in the microconstitution of natural flavouring substances and indeed of food itself. The analytical problem posed by the complexity and uncertainty of the available data means that using this method to demonstrate non-compliance with legislation is impractical. Even if flavour control were possible for nationally manufactured food, there would be no certain way of ensuring that imported food was in compliance.

LEGISLATION SYSTEMS

Having defined the terminology for flavouring ingredients, we should now look at available legislative systems:

(a) *Positive*. Only those materials that are listed may be used to the exclusion of all others. From the point of view of the legislator this system is ideal since it apparently produces total control. It puts the responsibility on the legislator for the safety of the products, without any way of being certain that the legislation is being adhered to.

(b) *Negative*. Any material may be used unless it is listed as unsuitable. This system is disliked by legislators because it does not offer theoretical control. However, if it is backed up by general legislation such as in the UK,[5] it offers a tenable alternative. The back-up legislation must make it an offence to add anything that is harmful to food. The responsibility is then firmly on the flavour manufacturer to guarantee the safety of his product, and since any proven example is likely to result in the demise of the offending company, you can be assured that this is taken extremely seriously.

(c) *Mixed system*. This really means a part positive and part negative system. The international flavour industry has accepted that a system which has artificial ingredients controlled by a positive (permitted) list and natural and nature identical ingredients controlled by a negative (forbidden or limited) list is a tenable compromise and has persuaded a number of national governments such as those of West Germany and Holland to their way of thinking.

THE IMPACT OF LEGISLATION SYSTEMS ON DEVELOPMENT

We must now consider examples of what has happened in countries where particular legislation has been in operation.

In the USA a positive list system has been in operation since 1965. This is known as the GRAS system, standing for 'generally regarded as safe'. The responsibility for clearance of flavouring materials is vested in the Flavour and Extract Manufacturers Association (FEMA) by the Food and Drug Administration (FDA) although it is possible to get materials cleared privately without publication. What has happened in this case? The original GRAS list of 1965 contained approximately 500 nature identical materials together with 370 naturals and about 250 artificials.[6] By 1984 the nature identicals had increased to about 1100 with a negligible increase in naturals and artificials (see Fig. 1). In the same period the number of published volatile materials present in food had increased from about 600 to almost 4500 (see Fig. 2),[7] and I estimate that at least a further 1000 are known to the flavour industry but have not been published. As an example of unpublished data, we in PFW have investigated cooked liver in some depth and have confirmed the presence of 180 compounds which do not appear in published data (see Table 1). In my opinion it is inconceivable that so many of the compounds discovered in natural foods are found to be of no interest to US flavour companies to improve the quality of their flavours. It is rather that the cost of providing the required toxicological data, for clearance, practically limits the number of compounds that can be added to the positive list. The present

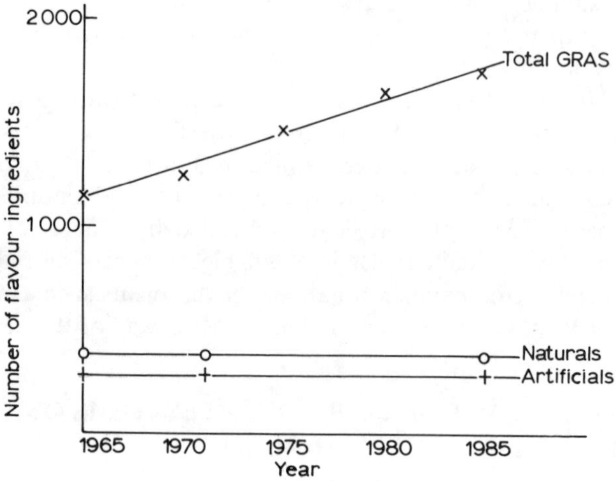

FIG. 1. Flavour ingredients generally regarded as safe in the USA (1965–85).

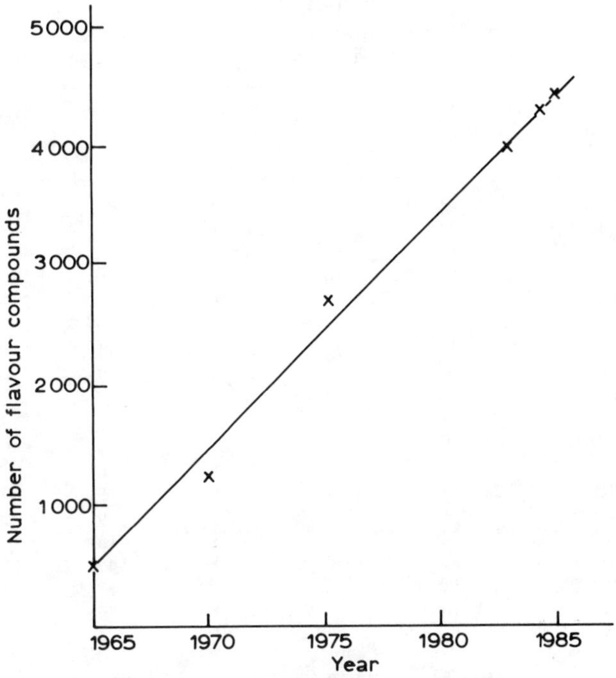

FIG. 2. Volatile compounds in food (1965–85) (by permission of Dr H. Maarse, VIVO Institute TNO, Zeist).

minimum cost for each clearance is estimated to be in excess of $50 000 per compound assuming that there are no complications.

From a development point of view, any traveller to the US will be struck by the crudeness of the flavourings present in those products which depend mainly on flavour for their acceptance, such as powdered drinks, sugar confectionery or dessert mixes.

What then has the US flavour industry done to counteract the constraints imposed by the GRAS list system? Any positive list system cannot effectively deal with non-defined ingredients such as process flavours, so these offer an alternative way of providing flavours which are outside the positive control. Added to this, the US food industry has demanded that more of the flavours that they purchase may be labelled as natural, so the US flavour industry has turned its attention and research resources towards the provision of non-defined natural flavours produced by techniques such as

TABLE 1
Liver Constituents: Distribution of Compounds Over Various Classes (PFW Unpublished Data)

Compound class	*Identified at PFW*	*New to liver*	*Previously published in liver*	*Total known in liver*	*New to food*
Hydrocarbons	24	17	29	46	0
Alcohols	22	14	15	29	0
Ethers	11	10	4	14	1
Aldehydes	31	10	29	39	2
Ketones	28	13	26	39	2
Acids	—	—	2	2	0
Esters	35	29	12	41	6
Lactones	—	—	2	2	0
Nitriles	2	2	2	4	0
Phenols	1	—	1	1	0
Furans	10	5	25	30	3
Pyrroles	1	1	5	6	0
Oxazoles	7	7	1	8	5
Thiophenes	11	7	12	19	6
Thiazoles	14	13	3	16	5
Other S-compds	39	30	15	45	11
Pyridines	15	15	—	15	8
Pyrazines	13	4	35	39	1
Amines	3	3	1	4	1
Total	267	180	219	399	51

heating together hydrolysed proteins, amino acids, sugars or microbiological or enzymic modification of natural substrates (usually food, but not exclusively).

In addition, flavours compounded of so-called natural fine chemicals have been developed. For instance 'natural' isoamyl acetate could be produced by allowing fermentation fusel oil (natural amyl alcohol) to stand, or be heated under 'kitchen' conditions with anhydrous vinegar (natural acetic acid). Whether a catalyst, natural or otherwise, is allowed is not very clear. Many 'natural' versions of important flavour fine chemicals are now being offered. They are identical in all normal respects to their 'synthetic' counterparts except they are usually 10–100 times the price. Obviously with this price differential there must be a temptation to the unscrupulous manufacturer to dilute the natural product with its synthetic counterpart and still label it as natural. To counteract this, US flavour manufacturers are becoming involved in the determination of naturalness using the C13/C12 stable isotope ratio method with all its attendant costs. As an example of the history of the dilution of naturals with synthetics, in 1983 my company was unable to purchase natural bitter almond oil which had not had synthetic benzaldehyde added. As a result of the US demand for naturalness, genuine bitter almond oil (100% natural) is now available but at a much higher price level.

Whilst it seems to be that research into process flavours or microbiological or enzymic flavours is very worthwhile, to try to provide natural fine chemicals, identical to their synthetic counterparts, is a total waste of resources and should not be used by food manufacturers to mislead the consumer into imagining that natural flavours containing them are somehow superior to their synthetic counterparts.

As a final comment on the US system, one should note that, when FDA Commissioners were asked how they could ensure that importations of food into the US contained flavours that conformed to their positive list, they admitted that it was not practicable.

Whilst it would be preferable to use an example of a country with a legislation based on entirely negative listing, such an example does not exist. The nearest which is available is that of West Germany where there exists a negative list for naturals and nature identicals and a very short list of artificials. Again the way that the flavour industry has developed in this market has been determined by the desire of the food industry not to have to label its products as artificially flavoured.

However, such labelling allows the use of both natural and nature identical flavouring ingredients without having to be labelled as artificial. In general this discrimination in favour of nature identicals has stimulated research into substitutes for commonly used artificials. This has resulted in looking into natural foods such as strawberry to find a suitable substitute for the phenyl glycidates which have traditionally been used in strawberry flavours. So far the nature identical strawberries that have been produced are not so characteristic of the fruit as their artificial counterpart.

In particular cases, widely used artificials are regarded as acceptable and ignored for labelling purposes. Ethyl vanillin is disregarded as an artificial if it is used in non-vanilla type flavourings, and ethyl maltol is regarded as a flavour enhancer and thus does not require labelling on most foods, and is disregarded as an ingredient of the flavouring.

However, the ready acceptability and wide definition of nature identity in West Germany has stimulated the international flavour industry into putting resources into research into the analysis of natural products used as food and the subsequent synthesis of novel naturally occurring synthetics. This is to the benefit of the flavour industry, the food industry and eventually the consumer, as it results in flavourings which are more representative of the natural foods which they imitate, both in character and constitution.

Finally, I would like to examine the present position in the UK. There is no specific flavour legislation but the use of all flavouring ingredients is within the requirements of the Food Act 1984.[5] This states that it is an offence to add to the food supply any material which is deleterious to health. It is interesting that this puts the total responsibility on the flavour supplier for the safety of the product. As must be obvious, this means that the supplier has to be confident of his position, since just one example of a company supplying an unsafe flavouring would probably finish the business for ever. However, the lack of specific legislation does allow for resources to be devoted to all forms of research into new flavourings, the main areas being new nature identicals, process flavours and microbiological and enzymic flavours. At the moment there is only one ingredients declaration required, viz. 'flavouring,' although this is sometimes qualified by the word 'natural'.[8]

In my view this adjective should be confined to flavourings derived solely from the named food, e.g. natural raspberry flavouring should be derived solely from raspberries, since I believe that this is what the consumer expects of such a flavouring and anything else would be regarded as misleading. Maybe it would be possible to introduce a class of natural flavourings where their ingredients were not totally derived from the named product, and I suggest that the term 'natural flavour blend', as used in the USA, might be appropriate. However, I emphasise that this is a personal view not agreed by the whole UK industry.

So, in summary, the flavour industry would have preferred either no specific legislation or legislation based on a negative list. As a fall-back position, we have accepted a positive list of artificials, and negative lists of naturals and nature identicals. This will benefit the consumer by improving the organoleptic quality of his food whilst exposing him to no greater risk than that posed by the food supply itself. We reject total positive listing as being against everyone's interest except that of the legislators.

REFERENCES

1. IOFI (1985). *Code of Practice for the Flavour Industry*, 2nd edn, C4 *et seq*.
2. STOFBERG, J. and STOFFELSMA, J. (1980). *Perf. Flav.*, **5**(7), 19. See also: STOFBERG, J., *Perf. Flav.*, (1981) **6**(4), 69; (1983) **8**(3), 61; STOFBERG, J. and KIRSHMAN, J. C. (1985). *Food Chem. Toxicol.*, **23**(9), 857.
3. BRITISH ESSENCE MANUFACTURERS ASSOCIATION (1982). Unpublished survey.
4. IOFI (1985). *Survey of World Usage of Artificial Flavouring Substances*, unpublished.
5. *Food Act 1984*, HMSO Publications.
6. HALL, R. L. and OSER, B. L., *Food Technol.*, (1965) **19**(2), 151; (1970) **24**(5), 25; OSER, B. L. and HALL, R. L., *ibid.* (1972) **26**(5), 35; OSER, B. L. and FORD, R. A., *ibid.*: (1973) **27**(1), 64; (1973) **27**(11), 56; (1974) **28**(9), 76; (1975) **29**(9), 70; (1977) **31**(1), 65; (1978) **32**(2), 60; (1979) **33**(7), 65; (1984) **38**(10), 65.
7. *Volatile Compounds in Food*, Division for Nutrition and Food, Research TNO, Institute CIVO-Analysis TNO, Zeist, Holland. Abstracted from editions dated 1963–85.
8. *SI 1305.1984*, *Food Labelling Regulations 1984*, HMSO Publications.

14

New Horizons in Toxicology and Nutrition

S. D. Gangolli

British Industrial Biological Research Association, Carshalton, Surrey, UK

ABSTRACT

The Council of Europe 'Blue Book' and the Food Additives and Contaminants Committee review contain about 1030 chemically defined synthetic flavouring substances. Of this list of flavours, 276 compounds have been considered provisionally acceptable for food use, but further toxicological information is required. The 'Blue Book' identifies a further 1300 flavouring substances, including about 500 natural ones, on which toxicological data are lacking to support their safety-in-use. Conventional studies for the safety evaluation of a compound are time-consuming and expensive, and resources, in terms of expert personnel and animal facilities, are limited, so there is an urgent need for alternative approaches to be explored. Cramer et al. *have developed a 'Decision Tree' scheme based on a sequence of 33 questions for the ranking of chemicals into three classes for the purpose of assigning priority in toxicity testing. This scheme has since been refined and updated and applied to a variety of chemicals of technological importance. The results obtained on applying this scheme to the flavouring compounds are presented. The scheme lends itself to further improvement by the incorporation of quantitative structure–activity relationship concepts, thereby providing a method for predicting the potential toxicity of a chemical and the likely target organ affected. The merits of such an approach are considered in the context of designing alternative* in vitro *and cell culture systems for investigating toxicity. The influence of nutrition in toxicology is highlighted in the context of the metabolic disposition and toxicity of ingested flavouring compounds.*

The safety evaluation of a food chemical in terms of human health is based on toxicity data in experimental animals, the likely level of the test chemical in the diet and the estimated human daily intake. In the United Kingdom animal data are obtained from conventional toxicity tests conducted according to the guidelines prescribed by the Ministry of Agriculture, Fisheries and Food in 1965.[1] In addition, guidelines for mutagenicity and carcinogenicity test procedures have been described by the Department of Health and Social Services.[2-4] The mutagenicity test procedures have been further updated by the United Kingdom Environmental Mutagen Society in their report of 1983.[5] Thus, the Committee on Toxicity of Chemicals in Food, Consumer Products and the Environment, responsible for advising the UK government, requires certain information, summarised in Table 1, for the safety evaluation of chemicals.

The facilities required for conducting the appropriate toxicity studies to fulfil regulatory requirements include sophisticated and expensive animal accommodation, specialised equipment and instrumentation and the collaborative efforts of specialists in the various scientific disciplines involved in toxicology. The cost of undertaking

TABLE 1
Information Considered Necessary for the Safety Evaluation of Food Chemicals (DHSS Report 27, 1982)[6]

1. Chemical structure, purity, formulation, processing details
2. Past and/or future likely extent of human exposure
3. Acute toxicity and target organ studies (up to 28 days studies)
4. Metabolism
5. Sub-acute toxicity (e.g. 90 days studies)
6. Long-term toxicity/carcinogenicity (e.g. rat study for 2 years minimum duration)
7. Reproduction/teratogenicity studies
8. Mutagenicity studies
9. Additional laboratory studies (e.g. potential allergenicity)
10. Human studies

Additional information requirements of the Food and Drugs Administration, USA[7]

1. *In utero* exposure to test compound
2. Multigeneration studies
3. Chronic studies extended to 20% survival of animals
4. All data generated from studies in compliance with Good Laboratory Practice (GLP) requirements

the full range of conventional toxicity studies on a food chemical at the present time is approximately £300 000 and the time required is 3–4 years. Thus, the substantial financial commitment necessary for obtaining the safety clearance of a food chemical is bound to be an important consideration in the economics of the development of new food additives and processes.

Furthermore, there has been increasing disquiet felt and reservations expressed by many eminent toxicologists about the validity and relevance of the mass of animal data requirements in assessing the toxic hazard of a chemical to man. The evaluation of the health hazard of a food chemical is based essentially on a 'no effect' dose level in experimental animals. By applying an arbitrary safety factor (usually 100 or 1000) the Acceptable Daily Intake (ADI) for man is derived. Clearly, this approach is of great regulatory convenience in that a permissible level of an additive or a permissible limit of a contaminant in food can be prescribed. This approach suffers from two defects. First, the arbitrary use of a safety factor in the extrapolation of animal data to man takes little account of species differences in pharmacokinetics, bioactivation, detoxication and repair mechanisms at target organ sites, and the modifying influence of age, nutrition and genetic factors on chemical toxicity. Secondly, the concept of a 'no effect' level is constantly being eroded by advances in biological sciences and analytical expertise. As toxicological methodologies become more sensitive, so more biological changes in treated experimental animals are recognised and the 'no effect' levels diminish. An inevitable consequence has been the increasing detection of biological effects at usage concentrations of food chemicals. Thus, on both economic and scientific grounds, the present prescribed procedures for the regulatory safety clearance of a food chemical cannot be viewed with complacency and this suggests a situation that leaves much to be desired.

Viewed in the light of the foregoing comments, the safety clearance of food flavours presents a formidable problem. The Council of Europe 'Blue Book'[8] and the FAC report[9] on the review of flavourings in food contain about 3000 chemically defined flavouring substances. The Council of Europe, on the advice of their expert committee, have categorised these flavouring substances into three lists. List I contains approximately 75 flavours which may be added to foodstuffs without hazard to public health. List II contains 276 flavours provisionally acceptable for use in food but about which further toxicological information is required. There is a third list containing 1800 flavours on which there is inadequate data to support their safety in use. In addition the 'Blue Book' lists about 500 natural flavours of plant or animal origin, substances of complex and varied chemical composition, on which relevant toxicological information is lacking. In view of the obvious logistic and economic problems associated with the use of conventional toxicity testing procedures for the safety evaluation of food flavouring substances, there is a need for alternative approaches to be explored.

The first approach worth considering is the Decision Tree scheme developed by Cramer *et al.*[10] for categorising chemicals into one of three classes for the purpose of assigning priority in toxicity testing. Class 1 chemicals are those with structures and related data suggesting a low order of oral toxicity and thus of low priority for toxicity testing. Class 3 chemicals are those that permit no strong presumption of safety and are therefore allocated a high priority for testing. Class 2 comprises chemicals about which there is inadequate information.

The format of the Decision Tree scheme of Cramer *et al.* consists of a sequence of 33 questions based on available information relating chemical structures and oral toxicity data. The main features of the scheme are shown in Table 2. The authors of the scheme evaluated 227 known carcinogens, all but one of which were found to fall into Class 3 (the exception being xanthine, a compound whose carcinogenicity is in doubt). However, the scheme has certain important limitations in that it cannot be applied to polymers and inorganic compounds. Additionally, some of the questions framed in somewhat ambiguous terms are difficult to answer in a simple yes/no form.

In the context of a MAFF sponsored research project for the development of an appropriate screening procedure for allocating toxicity testing priority to indirect food additives, BIBRA investigated the feasibility of applying the Decision Tree scheme. In order to facilitate the use of the Decision Tree, attention was given to redrafting a number of questions in a more explicit chemical form, and to extend the number of functional groups considered. An amended and extended scheme has been devised and found suitable for assessing the toxicity testing priority of a range of indirect food additives.[11]

The modified scheme has also been used for processing the

TABLE 2
Outline and Main Features of the Decision Tree Scheme[10]

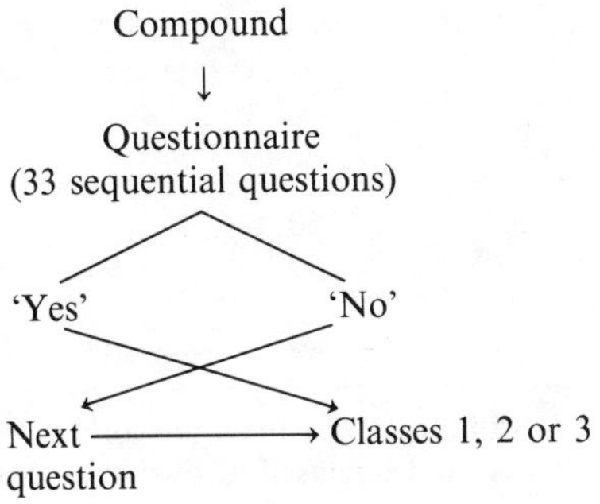

	Toxicity testing
Class 1–generally safe	low priority
Class 3–potentially toxic	high priority
Class 2–lack of information	don't know

Questionnaire: main features
1. Chemical structures include acyclic, alicyclic and aromatic types
2. Biological activity, chiefly carcinogenesis
3. Normal constituents of the body are considered to be Class 1
4. Does not cover inorganics and polymers

flavouring compounds in Lists I and II of the Council of Europe 'Blue Book'. The results, shown in Table 3, reveal that 209 flavouring substances could be considered in Class 3, i.e. in the category for high priority in toxicity testing. The majority of these substances belong to one of five chemical groups: allyl alcohol derivatives, anthranilates, eugenol derivatives, cinnamates and heterocyclic compounds. If the flavouring substances in Classes 2 and '0' were included, the total number requiring to be tested with a moderate to high degree of priority would be 335 out of a total of 1027 flavouring substances in Lists I and II (i.e. about 33%). This number is bound to be considerably increased if the flavouring substances in List III and the natural flavourings used in foods were to be assessed by the modified Decision Tree scheme. Plainly, the safety clearance of the flavouring substances by conventional procedures presents mammoth logistic problems and the expense involved would be grossly out of proportion to the costs of the minute levels at which many are used in foods. Thus there is an urgent need for alternative

TABLE 3
The Classification of List I and List II Flavouring Substances[8] According to the BIBRA Revised Decision Tree Scheme[10]

Flavouring substances	*Class*				*Total*
	1	*2*	*3*	*'0'*[a]	
List I	555	50	114	35	754
List II	137	39	95	2	273
	692	89	209	37	1027

[a] Class '0' consists of structural analogues of allyl alcohol and its oxidation products which were left unclassified owing to lack of biological data.

approaches to be developed in the context of the safety assessment not only of these flavouring substances and other food additives but also of indirect food additives, contaminants and environmental chemicals.

Two approaches that are receiving increasing attention and merit consideration are: (1) the use of quantitative structure–activity (QSAR) principles in predicting the biological activity of chemicals; (2) the development of alternative experimental techniques for complementing, or the replacement of, conventional animal studies. Each of these approaches will be briefly reviewed.

The correlation of biological properties with chemical structure has its roots in the study of the pharmacological action of drugs in experimental animals and man, and the assessment of the biological activity of analogous chemical structures has provided the necessary information for identifying functional chemical groups with pharmacological or toxic action, i.e. the pharmacophores or toxicophores.[12–14] The second major source of knowledge that has contributed to the development of a disciplined approach to quantitative structure–activity relationships is the role of metabolic pathways in mediating the bioactivation and detoxication of xenobiotics in mammalian systems (see reviews by e.g. Williams,[15] Parke,[16] Sies *et al.*[17]). The third vital component essential for the structuring of QSAR has been the correlation of the physico-chemical properties of chemicals with their biological disposition in mammalian systems. The properties that have been most widely used are shown in Table 4. The integration of these various strands has enabled the development of various QSAR systems. Examples of these systems

TABLE 4
Some of the Molecular Parameters Widely Employed in Quantitative Structure–Activity Relationships (QSAR)

1. *Hydrophobicity (lipophilicity)*: Hansch[18] or linear free energy method, log(partition coefficient octanol/water)
2. *Ionisation*: Hammett equation, σ-value
3. *Steric effects*: 1. Taft parameter (E_s)
   2. Molar refractivity (MR), Lorentz–Lorenz equation
   3. Parachor (P)
4. *Reduction potential*[19]
5. *Bond-dissociation energy*[20]

Other parameters: hydrogen bonding, infrared frequency, electron density, etc.

are the ones proposed by Tinker,[21] Enslein and Craig[22] and Chou and Jurs.[23] Recent developments in this topic have been reviewed by Golberg.[24] Other considerations known to modify absorption, biological disposition, metabolic activation, and detoxication processes and interactions with pharmacological receptor sites, and thereby to influence structure–activity relationships, are age, nutritional status and genetic factors. Thus, in the design of QSAR systems for predicting toxicity in an animal model and in the extrapolation of animal data to man, due regard needs to be taken of these factors. The influence of nutrition in toxicology is gaining increasing attention and recognition as evidenced by the burgeoning literature on the subject (see e.g. Parke and Ioannides,[25] Hatchcock and Coon,[26] Gershwin *et al.*[27]).

The development of alternative procedures as replacement for conventional animal toxicity studies has centred largely on the use of *in vitro* studies on cells in culture. Mammalian and bacterial cells in culture have long been of interest to those working on toxicological problems, for a number of reasons. *In vitro* systems offer the opportunity of studying complex problems under carefully controlled conditions and of dissecting the tangled and interrelated processes flowing from the interaction of xenobiotics and cellular constituents more thoroughly and efficiently than is possible in the intact animal. *In vitro* studies can be carried out rapidly and with relative ease, and changes in the cell culture microenvironment, to reflect changes in whole body physiology, can be readily effected. Furthermore, *in vitro* studies can be carried out on human tissues, tissue fractions or isolated primary cells in culture, thus providing information which may be more relevant to the estimation of human risk than data derived from experimental animals. In addition to extensive reviews on this subject, proceedings of conferences devoted to tissue culture applications in toxicology are an important source of recent development in this exciting new field. Thus, for example, the proceedings of the third international workshop on this subject have appeared,[28] and the proceedings of an international *in vitro* toxicology conference held in 1985 at the University of Reading are due to appear shortly.[29]

Clearly, before the QSAR techniques and alternative testing procedures can be applied for the safety evaluation of food flavouring substances, appropriate validation studies would have to be carried out on a range of model chemical toxins of known chemical structures and physico-chemical properties acting on specific target organs and on known intracellular systems. Doubtless, fragmentary bits of information germane to this important subject are published in the literature or available in the archives of research establishments actively involved in drug development and chemical toxicology. A concerted and systematic effort to collect and collate the relevant information from diverse sources, and the identification of the gaps in knowledge that require to be filled by conducting appropriate toxicological studies, would constitute an essential step on the route to the development of alternative systems for generating sound and meaningful toxicological data for assessing the health risk from chemicals to man.

REFERENCES

1. MAFF (1965). Memorandum, Ministry of Agriculture, Fisheries and Food, UK.
2. DHSS (1979). *Guidelines for Mutagenicity Test Procedures on Food Chemicals*, Department of Health and Social Security, UK.
3. DHSS (1981). *Guidelines for the Testing of Chemicals for Mutagenicity*, Department of Health and Social Security, UK.
4. DHSS (1982). *Guidelines for the Testing of Chemicals for Carcinogenicity*, Department of Health and Social Security, UK.
5. UKEMS (1983). *Report of the UKEMS Sub-Committee on Guidelines for Mutagenicity Testing*, Parts 1 and 2, United Kingdom Environmental Mutagen Society, UK.
6. DHSS (1982). *Guidelines for the Testing of Chemicals for Toxicity*, Department of Health and Social Security, UK.

7. FDA (1978). *Code for Good Laboratory Practice*, Food and Drugs Administration, USA.
8. COUNCIL OF EUROPE (1981). *Flavouring Substances and Natural Sources of Flavourings*, 3rd edn, Maisonneuve, Strasbourg.
9. MAFF (1976). *Food Additives and Contaminants Committee Report on the Review of Flavourings in Food*, Ministry of Agriculture, Fisheries and Food, UK.
10. CRAMER, G. M., FORD, R. A. and HALL, R. L. (1978). Estimation of toxic hazards: a Decision Tree approach, *Food Cosmet. Toxicol.*, **16**, 253.
11. PHILLIPS, J. C., PURCHASE, R., WATTS, P. and GANGOLLI, S. D. (1986). An evaluation of the Decision Tree approach for assessing priorities for safety testing of food additives, *Toxicologist*, **6**, 213 (Abst 858).
12. ARIËNS, E. J., SIMONS, A. M. and OFFERMEIER, J. (1978). *Introduction to General Toxicology*,
13. ARIËNS, E. J. (1980). In: *Drug Design*, Vol. 9, Academic Press, New York, 1–46.
14. ARIËNS, E. J. (1984). Domestication of chemistry by design of safer chemicals: structure–activity relationship, *Drug Metab. Rev.*, **15**(3), 425.
15. WILLIAMS, R. T. (1959). *Detoxication Mechanisms. The metabolism and detoxication of Drugs, Toxic Substances and other Organic Compounds*, 2nd edn, Chapman & Hall Ltd, London.
16. PARKE, D. V. (1976). The impact of drug metabolism on medical research. Part 1. Survey of metabolic processes, *Chem. and Ind.*, 380.
17. SIES, H., WENDEL, A. and BORS, W. (1982). In: *Metabolic Basis of Detoxication: Metabolism of Functional Groups*, Eds Jakoby, W. B., Bend, J. R. and Caldwell, J.
18. HANSCH, C. (1971). Quantitative structure–activity relationship in drug design. In: *Drug Design*, Vol. 1, Ed. Ariëns, E. J., Academic Press, New York, 271.
19. ADAMS, G. E., CLARKE, E. D., JACOBS, R. S., STRATFORD, I. J., WALLACE, R. G., WARDMAN, P. and WATTS, M. E. (1976). Mammalian cell toxicity of nitro-compounds: dependence upon reduction potential, *Biochem. Biophys. Res. Commun.*, **72**, 824.
20. SLATER, T. F. (1972). *Free Radical Mechanisms in Tissue Injury*, Pion, London.
21. TINKER, J. (1981). Relating mutagenicity to chemical structure, *Inf. Comput. Sci.*, **2**, 231.
22. ENSLEIN, K. and CRAIG, P. N. (1982). Carcinogenesis: a predictive structure–activity model, *J. Toxicol. Environ. Health*, **10**, 521.
23. CHOU, J. T. and JURS, P. C. (1979). Computer assisted structure–activity studies on chemical carcinogens: N-nitroso compounds, *J. Med. Chem.*, **22**, 792.
24. GOLBERG, L. (1983). *Structure-Activity Correlation as a Predictive Tool in Toxicology*, Hemisphere Publishing Corp., New York.
25. PARKE, D. V. and IOANNIDES, C. (1981). The role of nutrition in toxicology, *Ann. Rev. Nutr.*, **1**, 207.

26. HATCHCOCK, J. N. and COON, N. (1978). *Nutrition and Drug Interrelations*, Academic Press, New York.
27. GERSHWIN, M. E., BEACH, R. S. and HURLEY, L. S. (1983). *Nutrition and Immunity*, Academic Press, New York.
28. PROCEEDINGS OF THE THIRD INTERNATIONAL WORKSHOP ON TISSUE CULTURE APPLICATIONS IN TOXICOLOGY (1985). *Xenobiotica*, **15**, 635–779.
29. PROCEEDINGS OF THE INTERNATIONAL CONFERENCE ON PRACTICAL *in vitro* TOXICOLOGY (1986). *Food Chem. Toxicol.*, **24** (in press).

15

Food Off-flavours: Cause and Effect

F. B. WHITFIELD

CSIRO, Division of Food Research, North Ryde, New South Wales, Australia

ABSTRACT

Off-flavours in foods continue to be a major source of concern to the food industry, and of annoyance to the consumer. Even the introduction, in recent years, of strict quality controls in handling, processing and packaging has not completely eradicated all causes of off-flavours. The off-flavours most frequently described by consumers are disinfectant, mouldy, musty, earthy, catty, rotten cabbage, faecal, metallic, painty and plastic-like. These can be produced in the fresh food by microbial and chemical transformations and by chemical reactions during processing. They can also be introduced by the absorption of chemicals from the air, from the water used in processing, from packaging materials and from the environment in which the food has been grown, processed and stored. The chemicals responsible are principally volatile organic compounds which can be present at concentrations less than 1 part in 10^9. *Many of the compounds have been identified and analytical techniques developed for their detection and quantitation. Major problems arise from the variety of pathways by which the compounds can be incorporated into foods and how, with the relatively limited facilities available in many food factories, raw materials and the processed products can be efficiently monitored. This paper highlights a number of compounds known to be responsible for off-flavours, and describes some of the ways in which they can be incorporated into foods, and the effect that they can have on the acceptability of the contaminated product. It also describes how short-term solutions to off-flavour problems can be found once the responsible compounds have been characterised.*

INTRODUCTION

Every year the food industry receives complaints from consumers concerning off-flavours in fresh, processed and packaged foods. Fortunately, many of these are only isolated cases but, on occasions, major problems are encountered which may involve, for example, the destruction of a day's production or the condemnation of several freight container loads of packaged food. Some of the off-flavours most frequently encountered by consumers are described as disinfectant, mouldy, musty, earthy, faecal, metallic, fishy, painty and plastic-like.[1] In recent years consumers have become more aware of the occurrence of such off-flavours and are now more vocal in their criticism of those suppliers marketing inferior quality goods. The food industry in turn has become more sensitive to off-flavour complaints and now seeks an 'immediate' identification of the cause of their problems. However, the chemicals responsible for off-flavours are frequently present in foods at concentrations less than 1 part in 10^9 parts.[2] Furthermore, as the amount of contaminated material available for examination is usually small, the identification of such compounds, at least in the past, was a major challenge for the analytical chemist and sophisticated instrumental techniques. Fortunately, for all concerned, most of the compounds responsible for off-flavours are volatile organic compounds which, in most cases, makes them readily available for analysis by gas chromatography (GC) and combined gas chromatography and mass spectrometry (GC–MS).[2]

The commercial availability of ultra-high-resolution GC columns combined with the introduction of computer-controlled mass spectrometers, with the capability of mass chromatography and multiple ion monitoring mass spectrometry and the development of Fourier transform infrared and nuclear magnetic resonance spectroscopy, has simplified the chemist's task in recent years.[2] Thus, with the skilful use of instrumental and sensory techniques, a dedicated team of chemists can today characterise, if not identify, most compounds responsible for an off-flavour in a relatively short time—from a few days to several weeks. Identification of the source, or origin, of the off-flavour, and obtaining the evidence to convince the company involved to take a particular course of action, frequently a costly exercise, can be most time consuming. If we accept that off-flavour compounds can now be characterised with sufficient speed to almost meet industry's requirement of 'immediately', then the food industry is currently confronted by two major problems. These are the variety of pathways by which compounds can be incorporated into foods and how, with the relatively limited facilities available in many food factories, raw materials and the processed products can be efficiently monitored to prevent off-flavoured foods reaching the consumer.

The compounds responsible for off-flavours can be produced in the fresh food by microbial and chemical transformations,[1,3] and by chemical reactions during the processing and storage of food.[4,5] They can also be introduced into the food by the absorption of alien chemicals from the air, from the water used in processing, from packaging materials and from the environment in which the food has been grown, processed and stored.[1] It would be difficult for any food manufacturer or marketing outlet to monitor even the most common chemicals responsible for off-flavours, let alone check all of the various pathways by which these compounds could enter the food chain. Thus there will always be the occasional problem that will escape the quality control net. However, the more aware that food handlers and processors become of the types of compounds that can contaminate food, the effect that they can have on the acceptability of their product and the pathways by which these compounds can enter food, the better equipped the industry will be in controlling this type of problem. Off-flavour problems have no favourites and the company that escapes, for example, a batch of contaminated raw material could be the next casualty and suffer the same fate as its competitor. It is therefore essential that, as soon as an off-flavour problem has been identified, the details of the nature of the problem and its origin should be communicated to all food manufacturers and their ancillary industries. Such a commitment will not only benefit the food industry as a whole but also the consumer, by reducing unnecessary waste.

This paper will attempt to overcome this apparent lack of communication, throughout the industry, by highlighting a number of off-flavour compounds which have recently caused problems in the Australian food industry.

OFF-FLAVOURS ENCOUNTERED IN THE AUSTRALIAN FOOD INDUSTRY 1980–1985

Some off-flavour problems investigated at the CSIRO's Division of Food Research, North Ryde, over the past five years are listed in

TABLE 1
Off-flavours Encountered in Some Australian Foods 1980–5

Description	*Number of examples*	*Compounds responsible*	*Types of food involved*
Disinfectant	9	Chlorophenols	Soft drinks, flour, cocoa powder, canned vegetables, fruit juice, bulk chocolate, packaged biscuits, canned fruit, poultry
Musty/mouldy	5	Chloroanisoles	Wine, packaged flour, cocoa powder, chilled meat, dried fruit
Earthy/muddy	3	2-Methylisoborneol	Canned mushrooms, bottled wine
		Geosmin	Chilled meat
Garlic-like	2	Trimethylarsine	Prawns, sand-lobsters
		Bis(methylthio)methane	
Onion-like	1	Dimethyl trisulphide	Prawns
Metallic	1	Octa-1,5-dien-3-ol	Prawns
Faecal	1	Skatole	Potato chips, potato flakes
		Indole	
		p-Cresol	
Catty	1	Unknown (not 4-mercapto-4-methylpentan-2-one)	Chocolate
Animal/musty	1	Unknown	Bread
Cabbage-like	1	Dimethyl sulphide	Peanuts
Medicinal	1	Iodoform	Canned beer
Kerosene	1	*p*-Cymene	Cordials
Iodoform-like	1	Bromophenols	Prawns

Table 1. Of these twenty-eight problems, nine involved contamination of food with chlorophenols and five involved chloroanisoles. In general, the chlorophenols are the precursors of the chloroanisoles;[1] thus chlorophenols, as such or as their derivatives, were involved in half the problems investigated. Not only were these two classes of compounds responsible for the greatest number of off-flavour problems but they were also accountable for the industry's most costly losses, principally expressed as damage to reputation and destruction of the contaminated product. The contamination of foods by chlorophenols and chloroanisoles provides examples of alien chemicals entering the food chain.

The next most frequent problems concerned the presence of earthy/muddy and garlic-like off-flavours in processed and fresh foods respectively.[3] Earthy off-flavours are frequently encountered in fresh and processed foods; however, in many of these cases, especially with soil-grown foods, it is difficult to distinguish between an off-flavour caused by contamination and that resulting from an intensification of a naturally occurring flavour component. In the three examples listed in Table 1 it was possible to show that the off-flavour was due to contamination.

Many off-flavour problems are sporadic, as is the case with most chlorophenol contaminations; yet, with the garlic-like off-flavours in certain species of crustacea, the absence of this off-flavour was considered to be the exception rather than the rule.[3] The industry, but not the consumer, came to accept this off-flavour as a natural flavour and was willing to sell this species at about half the price received for premium grade species. Research showed that, with improved handling techniques, formation of the off-flavour compounds could be prevented.

Single sources of contamination were responsible for the remaining nine problems listed in Table 1. The fact that each of these off-flavour problems was only associated with one complaint does not make the responsible compounds a less serious threat to the food industry. This point will be emphasised throughout the following discussion.

Disinfectant Off-Flavours

Of the nine problems highlighted in Table 2, four were the result of accidental contamination of food by chlorophenol-based herbicides and biocides during the processing and handling of the

TABLE 2
Cause and Origin of Disinfectant Off-flavours

Food contaminated	*Compounds involved*	*Origin*
Poultry	2,4,6,-Trichlorophenol 2,3,4,6-Tetrachlorophenol	Cage litter containing wood shavings contaminated with wood preservatives
	6-Chloro-2-methylphenol 2,4,6-Trichlorophenol	Chlorophenol-based biocides used to sterilise cages
Canned carbonated beverages	6-Chloro-2-methylphenol	Empty drink cans contaminated with airborne agro-chemical pollutants
Canned vegetables	2,4,6-Trichlorophenol 2,3,4,6-Tetrachlorophenol Pentachlorophenol	Process water contaminated with chlorophenol-based algaecide
Bulk chocolate	2,4,6-Trichlorophenol 2,3,4,6-Tetrachlorophenol Pentachlorophenol	Wooden pallets treated with a chlorophenol-based wood preservative
Modified flour	2,4-Dichlorophenol 2,4,6-Trichlorophenol	Wash water contaminated with in-plant chlorination products
Reconstituted fruit juice	2,4-Dichlorophenol 2,6-Dichlorophenol	Process water contaminated with in-plant chlorination products
Canned fruit	2,4,6-Trichlorophenol	In-plant reaction involving steam contaminated with phenol, and process water containing chlorine
Cocoa powder	2,4-Dichlorophenol 2,4,6-Trichlorophenol 2,3,4,6-Tetrachlorophenol	Multi-walled paper sacks contaminated with fungicide and waste-paper residues
Packaged biscuits	2,4,6-Trichlorophenol 2,3,4,6-Tetrachlorophenol	Adhesive containing chlorophenols as a preservative

product. Only one of these sources, the tainting of poultry by chlorophenol derivatives present in cage litter, has been adequately documented,[6,7] and in these cases absorbed chloroanisoles, and not chlorophenols, were responsible for the off-flavour. Even so, the company concerned should have been aware that the use of wood shavings treated with chlorophenols eventually would lead to some form of contamination in their product. The use, by the same company, of a chlorophenol-based bactericide would therefore appear most unwise. Although some of the ingredients of such preparations will be listed by the manufacturer, minor components, and certainly possible impurities, will not. In this case it was only after the problem had developed and the components of the cages had been analysed for chlorophenols that it was established that the bactericide was the source of the contamination. The supplier claimed that he was unaware that the product contained chlorophenols and, as the preparation was made up from components supplied only with trade names, this statement was probably true. Clearly, before using biocides in sensitive areas it is prudent for food manufacturers to have such materials analysed to establish that they are free from components such as chlorophenols.

The tainting of a carbonated beverage by an airborne pollutant typifies one of the most difficult forms of contamination that can confront a food manufacturer.[1] This form of contamination was first highlighted by Griffiths and Land,[8] who showed that 6-chloro-2-methylphenol, originating from an agrochemicals plant, was responsible for imparting an off-flavour to biscuits being cooled on the processing line. Even with this background information, the beverage company could be excused for not readily establishing the source of their problem. The plant had experienced chlorophenol problems on several occasions and these had all been traced to the reaction of trace quantities of phenol with chlorine water on the processing line. When it was established that this form of contamination was not occurring, the search was broadened. Eventually it was established that empty cans, incompletely covered by cardboard spaces, when stacked on pallets often gave an off-flavoured product. On examination of the warehouse where some of the cans were stored during off-peak periods, it was observed that the building smelled of chlorophenols and that an agrochemicals plant was located nearby. Furthermore, it had been observed that the off-flavour problem only occurred when the cans were stored at the warehouse during the spring and early summer months, the time of year when the wind blew preferentially from the direction of the agrochemicals plant. At the time of the visit to the warehouse the agrochemicals plant was using 4-chloro-2-methylphenoxyacetic acid, a herbicide frequently contaminated with 6-chloro-2-methylphenol, in its formulations. Obviously, before siting food plants and warehouses, the local industrial environment must be taken into consideration and, where chemical industries are situated nearby, precautions should be taken to prevent the absorption of odorous chemicals by food and packaging materials.

Algaecides are frequently used to prevent the growth of aquatic organisms in refrigerator condenser water, and under normal conditions this additive will never enter the processing line. Yet, when several batches of canned vegetables were found to be contaminated with the same chlorophenols as were present in the algaecide, it appeared likely that the condenser water was the likely source of the contamination.[1] In-plant investigations showed that, as a result of a sudden loss of water pressure, coupled with a series of interrelated mechanical faults, a small quantity of the treated water had been sucked back into the water supply to the processing line. When chemical treatments are being used in food processing plants, it is essential that treated water is completely isolated from the potable water supply, preferably by the use of header tanks; as shown in the above example, a system of non-return valves can fail.

Wooden pallets treated with chlorophenol preservatives have long been recognised as a source of off-flavours in non-hermetically sealed foods. But in Australia, where hardwoods are normally used for the construction of these items, and such timber need not be treated with preservatives, pallets are not a common source of off-flavour problems. Pallets treated with chlorophenols were shown to be the source of contamination in samples of bulk chocolate affected by an intense disinfectant off-flavour.[1] This problem could have been prevented by the use of heavy duty aluminium foil or, less preferably, thick fibreboard, to insulate the food from the timber. Thus, where sensitive materials are to be transported or stored on wooden pallets, it is prudent for food manufacturers and handlers to assume that the timber has been treated and to take precautions.

Another common cause of chlorophenol contamination of food results from in-plant chlorination of phenol.[1] The major source of chlorine in such cases is the treated water supply. Chlorine-based sterilising agents are probably the next most common source of this chemical. There are, however, a multitude of sources of trace quantities of phenol in food processing plants. Plastic and resin fittings and paints are a common source of this compound but phenol can also be derived from the food itself, from printing inks and packaging materials, and even from boiler water additives.[1]

In the problem involving batches of modified flour, the first sign that the company had a problem was the receipt of complaints from manufacturers of meat pies and batter coated foods.[9] It was soon established that the flour was the source of the off-flavour, but determining the origin of the contamination in the flour mill was more difficult. Since the chlorophenols involved were those usually associated with in-plant chlorination, the processing line appeared to be the most likely source of the contamination. Attempts to reproduce the problem under controlled conditions were unsuccessful. A complete examination of the plant followed before the chlorophenol source was traced to a storage tank containing sodium hypochlorite sterilising solution. This solution was heavily contaminated with chlorophenols, and examination of the tank, after its destruction, showed that the fibreglass lining had almost completely disintegrated. The source of the phenol was traced to the breakdown of the bonding resin of the fibreglass. Examination of plant records showed that the tainting problem had occurred soon after an outbreak of microbial contamination on the processing line. The hypochlorite solution, containing the chlorophenols, had been used to sterilise the infected processing equipment. It appeared likely that some of this solution had been left in the connecting pipes and that the first batch of flour on the processing line had been contaminated with the chlorophenols during the initial water wash.

Far more difficult to establish was the origin of the chlorophenol present in some batches of canned fruit.[1] Subsequent in-plant studies showed that the only products affected by the off-flavour were those which were subjected to live steam injection during the pre-process sterilisation processes. A closer examination of the process showed that a phosphate chemical, which contained lignin sulphonate, was routinely added to the boiler water to control hardness. Lignin sulphonate is a non-homogeneous product from the sulphite liquor wastes of paper mills; its major component is a complex phenolic compound which may give rise to polyphenols on heating at high pH but, more important, it may contain, as components, simpler phenols which yield phenol itself under these conditions. Laboratory experiments showed that, when the phosphate additive was autoclaved for 1 hour at pH 10·5 to 11, the steam distillate produced a chlorophenol-type off-flavour when treated with water containing 8–10 mg/l of chlorine. 2,4- and 2,6-Dichlorophenol and 2,4,6-trichlorophenol were detected in this solution and phenol in the initial steam distillate, providing the link between the lignin sulphonate and the off-flavour compounds.

The water supplies used in most food plants are chlorinated, some more heavily than others, but when chlorine and chlorine-based sterilising agents are being used in such plants it is essential that these reagents are kept well clear of any materials which can contain phenol, or can yield phenol under normal plant operating conditions.[1] It is therefore critical that engineers know exactly what materials are being used in their factories if the processing line is to be kept free of this form of contamination.

Technical grade sodium pentachlorophenate, which frequently contains as impurities 2,4,6-trichlorophenol and 2,3,4,6-tetrachlorophenol, has been used extensively in starch and animal protein based glues to provide antimicrobial protection during their manufacture, storage and service life.[10] Furthermore, many publishers use this fungicide to provide similar protection for their newsprint.[11] Chlorophenols are also produced during the bleaching of wood pulp for paper manufacture and can remain as residues in the finished product.[12] In high quality fibreboard, the concentrations of these chlorophenols are usually too low to cause contamination in food packaged in boxes manufactured from such materials.[11] However, regenerated fibreboard prepared from waste products, including newsprint, recycled fibreboard boxes and paper sacks, can have relatively high concentrations of chlorophenols.[11] Packaging materials manufactured from these regenerated materials can be a source of disinfectant off-flavours in non-hermetically sealed foods.

With the two examples of this type of contamination, listed in Table 2, chlorophenols derived from both the fibreboard packaging materials and the adhesives were shown to be the source of the off-flavour in the packaged food. In the case of the cocoa powder, 2,4-dichlorophenol was found to be the compound principally responsible for the off-flavour.[13] Four other chlorophenols were also present in the powder but only 2,4-dichlorophenol was present in concentrations which could be demonstrated to produce the off-flavour. Analysis of the paper sacking used to package the powder showed that, although the glued side-seams contained very high concentrations of pentachlorophenol, the paper was the major source of 2,4-dichlorophenol. This compound along with 2,4,6-trichlorophenol and 2,3,4,6-tetrachlorophenol are the principal chlorophenols formed during the chlorine bleaching of wood pulp for paper manufacture.[12] Furthermore, as microscopic examination of the sacking material showed that it mainly consisted of virgin fibres, the logical source of the contamination appeared to be the initial bleaching of the wood pulp.

Studies involved with problems in packaged biscuits showed that only the first few biscuits at the end of each packet were affected by the off-flavour.[1] Significantly, the packets were stacked on-end in fibreboard boxes. Analyses of the boxes showed that the concentration of the offending chlorophenols was much higher in the lids and bases than in the sides. Although the fibreboard in the boxes was prepared from recycled waste, and therefore had a relatively high basic chlorophenol content, the fact that only the lids and bases were associated with the problem indicated that the adhesive used in the final assembly was the principal source of the contamination.

In recent years in Australia there has been increasing pressure from food manufacturers that fibreboard and paper to be used in the packaging of non-hermetically sealed foods should be prepared from virgin fibreboard of low chlorophenol content.[11] Furthermore, some sections of the industry now require that adhesives destined for use in the manufacture and sealing of packaging materials should be free from chlorophenol preservatives.

Musty/Mouldy Off-flavours

In four of the five problems highlighted in Table 3 it was possible to demonstrate that the production of the offending chloroanisoles, from residual chlorophenols, was probably biological. In these cases several species of mould, with the ability to methylate phenolic compounds, were isolated from the contaminated packaging materials. In only one example was it not possible to isolate moulds which could carry out the above transformation. This involved the tainting of bottled wine by cork stoppers containing traces of chloroanisoles. The failure to isolate the moulds was not surprising, as the corks

TABLE 3
Cause and Origin of Mouldy/Musty Off-flavours[a]

Foods Contaminated	*Compounds involved*	*Origin*
Wine	2,4,6-Trichloroanisole	Corks initially contaminated with chlorophenols formed during chlorine bleaching process
Packaged flour	2,4,6-Trichloroanisole 2,3,4,6-Tetrachloroanisole	Jute sacks initially contaminated with chlorophenols present either as preservatives or formed by an in-plant chlorination reaction
Cocoa powder	2,6-Dichloroanisole 2,4-Dichloroanisole 2,4,6-Trichloroanisole 2,3,4,6-Tetrachloroanisole	Multi-walled paper sacks initially contaminated with chlorophenol residues
Chilled meat	2,4,6-Trichloroanisole	Wooden insulation in cavity walls of chiller cold-store initially contaminated with chlorophenol reaction products or residues
Packaged dried fruit	2,4,6-Trichloroanisole 2,3,4,6-Tetrachloroanisole	Fibreboard packaging material initially contaminated with chlorophenol residues

[a] The formation of the chloroanisoles in these fibrous materials was probably biological as several species of mould with the ability to methylate chlorophenols were isolated from each of these materials.

had been sterilised with sulphur dioxide before their insertion in the bottles. In the remaining examples, the packaging materials had received no such treatment either before or after the contamination of the food.

Jute sacks used for the packaging of various cereal grains and flours are known occasionally to possess unpleasant odours which can be imparted to the food on storage.[14] When some batches of Australian flour were found to possess a musty/mouldy odour, the source of the problem was speedily traced to several shipments of jute sacks; however, establishing the cause of the taint and the origin of the chemicals responsible proved more difficult. Eventually the compounds responsible were shown to be 2,4,6-trichloroanisole and 2,3,4,6-tetrachloroanisole, and these compounds together with the corresponding chlorophenols were also shown to be present in the contaminated sacks.[15] From these sacks were also isolated three species of mould, *Penicillium crustosum*, *Talaromyces flavus* and *Paecilomyces variotii*, which rapidly methylated 2,4,6-trichlorophenol in Czapek broth. The origin of the chlorophenols in the contaminated sacks was uncertain but there appeared to be two possible sources. Chlorophenols had been recommended for the rot-proofing of jute fibres,[16] and jute fabrics are bleached with chlorine[17] which could yield the chlorophenols isolated from the tainted sacks. Attempts to obtain supporting evidence from the manufacturers of the sacks were not forthcoming.

The compound responsible for the musty/mouldy off-flavour in the packaged cocoa powder was 2,4,6-trichloroanisole and this compound together with four other chloroanisoles were shown to be present in the glued side seams of the packaging material.[13] Also identified in the seams were the corresponding chlorophenols and five species of mould which readily methylated chlorophenols in synthetic media. Of these, three were *Penicillium* species and one each of *Mucor* and *Paecilomyces*. Conditions of high water activity and temperatures between 25°C and 35°C favour the growth of these moulds and, as the packaged powder originated in tropical Asia, it was believed that conditions such as these could have been experienced either before or during transport to Australia. The chloroanisoles so formed would then have been absorbed by the food during storage in local warehouses.

Quite a different source of chloroanisole contamination was encountered in an investigation of a mouldy off-flavour in meat exported from Australia.[1,18] The uncooked meat had no distinguishing aroma; however, after frying, the meat had a characteristic mouldy aroma and was quite inedible. Analysis of the meat showed that it was contaminated with trace quantities of 2,4,6-trichloroanisole. Determining the origin of this compound proved a difficult task. It was first traced to a block of four chillers in one meatworks. These chillers had a mouldy aroma and the chloroanisole was present in condensate on the refrigeration coils. Condensate was observed to drip occasionally onto the carcasses where the off-flavour compound was readily absorbed into the fatty tissue. A false ceiling was built to overcome this source of contamination. Although the intensity of the off-flavour decreased, the problem still existed since carcasses held in these chillers for extended periods had an intense mouldy off-flavour. Analyses of the atmospheres taken from within the chillers and immediately outside them indicated that the chloroanisole was present in both areas in minute concentrations. The insulation within the walls of the chillers was eventually identified as the likely source of contamination since this material was heavily contaminated with 2,4,6-trichloroanisole. It was therefore suggested that water, which periodically flooded the cavity walls, transported the chloroanisole through the relatively porous brick walls to the floor of the cold store, situated below the chillers, where it accumulated in the ice. Once the compound had reached the outside wall of the chillers, it was easily picked up by the air stream and circulated through the whole area where it was readily absorbed by the hanging carcasses. The formation of the off-flavour compound is uncertain. It is possible that it was produced in a two-step synthesis involving inadvertent chlorination of phenol, present as a minor component in the insulation material, by chlorine used in the sterilisation of the chiller rooms and corridors; the product 2,4,6-trichlorophenol then underwent fungal methylation to give 2,4,6-trichloroanisole. Supporting this hypothesis, a mixed culture of moulds isolated from the insulation material readily methylated 2,4,6-trichlorophenol in a synthetic medium.

With the final example, listed in Table 3, dried fruit exported from Australia was found occasionally to possess a musty off-flavour; the cause was contamination with traces of 2,4,6-trichloroanisole and 2,3,4,6-tetrachloroanisole.[11] The fruit had been packaged in fibreboard boxes and the product shipped overseas in metal freight containers. Only a few boxes per container were affected and only a small proportion of the shipped container loads developed the problem. No complaints were received concerning fruit packaged at the same time but sold in Australia. Analyses of the fibreboard boxes showed that they were contaminated with the offending chloroanisoles and pentachloroanisole. The boxes also contained high concentrations of the relative chlorophenol precursors, together with a collection of xerophilic moulds which readily methylated 2,4,6-trichlorophenol in Czapek broth. These moulds included *Scopulariopsis brevicaulis*, *Paecilomyces variotii*, three species of *Penicillium* and one each of *Eurotium* and *Aspergillus*. When samples of regenerated fibreboard containing high concentrations of chlorophenols were inoculated with mixed cultures of these moulds and stored at water activities of 0·75, 0·84 and 0·90 for 6 weeks at 25°C, subsequent analyses showed that the concentration of 2,4,6-trichloroanisole increased dramatically with increase in water activity. Mustiness in dried fruit first became a problem at the time of the introduction of containerisation. This form of transportation can produce major shifts in moisture levels within the cargo, with higher levels at the top of the container. The greater concentrations of chloroanisoles observed in the lids of the boxes agreed with the key role of moisture in the formation of these compounds. With regard to the high concentrations of chlorophenols found in the regenerated fibreboard boxes, the principal origin appeared to be local newsprint which made up the bulk of the recovered waste used in the recycling process.

These investigations highlight the problems associated with the occurrence of trace quantities of chlorophenols, particularly 2,4,6-trichlorophenol and 2,3,4,6-tetrachlorophenol, in materials which come into contact with non-hermetically sealed foods. In the presence of moisture these compounds may be converted into strongly flavoured chloroanisoles which all readily migrate into foods. To prevent the continued occurrence of such problems it is essential that the packaging and transport industries reduce the concentrations of chlorophenols in the environments in which foods are stored. Where biocides need to be used, they should be completely free from chlorophenols and their derivatives, and phenol and chlorine should never be used simultaneously for the sterilisation of plant and warehouse.

TABLE 4
Cause and Origin of Off-flavours Formed by Micro-organisms

Food contaminated	*Compounds involved*	*Origin*
Canned mushrooms	2-Methylisoborneol Geosmin	Origin not proved, but process water derived from either a reservoir or a storage tank contaminated with actinomycetes and/or cyanobacteria was believed to be responsible
Chilled meat	Geosmin	Wooden insulation in cavity walls of chiller cold-store contaminated with species of actinomycetes
Wine	2-Methylisoborneol Geosmin	Cork stoppers contaminated with both compounds but no micro-organisms were isolated
Prawns Sand lobsters	Trimethylarsine Bis(methylthio)methane	Origin not proved but studies using γ-irradiation supported view that offending compounds were formed by micro-organisms
Prawns[19]	Dimethyl trisulphide	Prawns contaminated with two species of *Pseudomonas*
Prawns[20]	2,4,6-Tribromophenol A dibromophenol	Likely origin marine algae
Potato chips and flakes	Skatole, indole and *p*-cresol	Potato tubers contaminated with several species of bacteria including two species of *Clostridia*

Off-flavours of Microbial Origin

Off-flavours are frequently caused by the introduction of microbial metabolites into food either by direct microbial action or indirectly by absorption from a contaminated environment.[3] Many of these metabolites remain in the food even after extensive processing. Where direct microbial action is involved, the food is frequently found to contain only low levels of microbial contamination, making the identification of the organism involved difficult, if not impossible. Furthermore, the major concern of food microbiologists is to ensure that foods under their control are microbiologically safe for consumption, and hence there is a certain unwillingness to search for an organism, not apparently a health risk, that may be responsible for an off-flavour. To demonstrate the need for greater involvement by microbiologists in the study of off-flavours, five recently investigated problems, listed in Table 4, believed to be of microbial origin, will be described.

With the problem involving imported canned mushrooms with a muddy off-flavour, 2-methylisoborneol was quickly established as the compound responsible for the taint.[21] Geosmin, a compound with an intense earthy aroma and flavour, was also identified in the tainted material but was present in concentrations well below its taste threshold. These compounds are produced by a number of species of soil organisms, the actinomycetes, and aquatic organisms, the cyanobacteria, and are frequently associated with objectionable tastes and odours in drinking water.[22] Significantly, the off-flavoured mushrooms were restricted to cans of the same batch number and, of the cans with this batch number, only 10% were affected. Examination of the brine and mushrooms showed that they were microbially sterile. Subsequent studies showed that it was not possible to induce 2-methylisoborneol into fresh mushrooms by the inoculation with actinomycetes of freshly picked mushrooms, mushroom spawn and compost used for mushroom cultivation. 2-Methylisoborneol occurring in municipal water supplies was, however, a possible cause of muddy off-flavours in foods canned in brine or syrup. The sporadic nature of the mushroom problem could have resulted from the build-up of this compound in feeder tanks and in connecting water pipes during an overnight or weekend shutdown. Under these conditions, only those cans processed first would have contained enough of the compound to produce an off-flavour. The added knowledge that, in the country where the product originated, some reservoirs were occasionally affected by blooms of actinomycetes and/or cyanobacteria offered some support for this hypothesis.

In addition to 2,4,6-trichloroanisole some samples of chilled meat affected by a musty off-flavour were found to contain significant concentrations of geosmin. Panellists agreed that in these samples the flavour of geosmin played a major role in the perceived off-flavour. The origin of the geosmin, like the 2,4,6-trichloroanisole, was the insulation in the cavity walls of the chiller cold store. Geosmin along with the chloroanisole was absorbed by the carcasses from the contaminated stores atmosphere. Examination of the insulation material showed that it contained several species of *Streptomyces* which, when cultured in synthetic media, produced the characteristic aroma of geosmin.

When unusual earthy or muddy off-flavours occur in fresh or processed foods, a search for possible sources of contamination from actinomycetes and/or cyanobacteria should be one of the first steps considered. In particular, water supplies should be investigated as a possible source of contamination if local reservoirs suffer occasional blooms of these organisms.

In the study of garlic-like off-flavours in crustacea, bis(methylthio)-methane[23] and trimethylarsine[24] were shown to be the compounds responsible for the unacceptable flavour of the product. Deep-sea prawns are particularly susceptible. These crustacea are usually soft-shelled and are easily damaged on handling, making them predisposed to microbial contamination. However, studies showed that the off-flavour was present in the catch as soon as it was landed on deck.[25] Because of the depth at which these prawns are caught (685 m), some of the catch would be held in the trawling net for up to 2 hours. If the prawns died soon after netting, it was considered possible that the off-flavour compounds could be formed while the catch was still on the ocean floor. Accordingly it was proposed that bis(methylthio)methane could be formed by the reaction of methanethiol, a bacterial metabolite, with formaldehyde, a decomposition product of trimethylamine oxide.[3] Methanethiol is known to be produced in sea foods by several species of *Pseudomonas*,[26] and formaldehyde has been shown to be rapidly produced in dead prawns.[27] Yet, attempts to relate specific bacteria to the production of bis(methylthio)methane proved unsuccessful. Similarly trimethyl-amine can be produced by a number of micro-organisms, including a soil mould, a yeast from sewage sludge, and *Methanobacterium* MoH, a bacterium isolated from ocean mud.[28] Crustacea accumulate arsenic in their tissues, and it was possible that a *Methanobacterium* in the digestive system of the prawn could produce trimethylarsine from accumulated inorganic arsenic derivatives.[3] Attempts to isolate the organism responsible for the transformation were unsuccessful.

Bis(methylthio)methane was also shown to be responsible for the garlic off-flavour in sand lobsters, and it is interesting to note that this problem only develops in the males.[3] This crustacean is caught in shallow water and is usually transported to the markets alive. But if a male lobster dies the flesh rapidly develops an intense garlic off-odour which contaminates the rest of the catch. However, it was observed that when freshly killed male lobsters were irradiated with gamma-rays, at levels known to destroy all non-spore-forming bacteria, the resulting material could be stored for 16 hours at room temperature without production of bis(methylthio)methane.[29] Once again, the organism responsible for the production of bis(methylthio)-methane was not identified.

When samples of frozen potato chips were found to possess an intense pigsty-like off-flavour, production of this product and other potato lines ceased and all suspect materials were recalled from retail outlets. Eventually, the compounds responsible for the off-flavour were shown to be *p*-cresol, indole and skatole;[30] with these identifications, in-plant studies were undertaken to establish their origins. Examination of potato tubers in storage sheds at the factory producing the potato products revealed that those tubers in contact with the colder southern wall exhibited malodorous soft rot and contained large quantities of the offending compounds. In addition, tubers not affected but stored adjacent to the malodorous material were also contaminated to varying degrees with *p*-cresol, indole and skatole. Eventually, a small proportion of freshly dug tubers showing signs of soft rot were also found to be contaminated with the same compounds. In Australia, soft rots in potato tubers result from an initial infection of injured tubers by species of *Erwinia*. However, species of *Clostridia* have also been implicated in the production of soft rots in potatoes, and some species of these bacteria are known producers of *p*-cresol, indole and skatole.[31] Inoculation of sterile potato cores with individual cultures of aerobic and anaerobic isolates from tubers affected by malodorous soft rots led to the development of soft rots with odours similar to those present in the

TABLE 5
Cause and Origin of Off-flavours Formed by Chemical Reactions

Food contaminated	*Compounds involved*	*Origin*
Prawns[32]	Octa-1,5-dien-3-ol	Product of copper-catalysed oxidation of polyunsaturated fatty acids
Chocolate	Unknown	Product formed in wrapping film by the reaction of alkyl mercaptans with a titanium complex of pentane-2,4-dione
Bread	Unknown	Compound, a trace component, in stearic acid used to prepare the food additive sodium stearyl lactilate
Peanuts[33]	Dimethyl sulphide	Product formed by the reaction of methyl bromide fumigant with methionine followed by thermal decomposition of resultant *S*-methionine sulphonium bromide
Beer	Iodoform	Product formed by reaction of an iodine-based sterilising agent with an alkaline solution containing ethanol
Kerosene	*p*-Cymene	Acid-catalysed transformation of citral to *p*-cymene

worst examples of off-flavoured chips. In the two inoculates resulting in the production of these rots, the bacteria were identified as species of *Clostridia.* Subsequent field studies showed that the initially infected tubers appeared to originate only in one or two growing areas. Here some farmers, faced with high soil temperatures during the harvest season, were found to water their paddocks heavily in order to lower temperatures and prevent physiological damage to tubers. Unfortunately, the bacteria that caused the soft rot thrived under these conditions.

In the investigation of off-flavour problems of possible microbial origin, the identification of the compounds responsible is the first important step, as this knowledge can be used to predict the type of organisms involved in the contamination. Such information is of great assistance to the microbiologist attempting the isolation. For example, with the contamination of the potato products by *p*-cresol, indole and skatole, the search was directed towards species of *Clostridia* as the causative organisms. With the classification of the organism its source can be more readily identified and industry can, in turn, introduce specific processes to prevent its spread and growth by improving handling procedures and by establishing higher standards of hygiene.

Off-flavours of Chemical Origin

Of the six problems listed in Table 5, the compounds responsible for two of the off-flavours have not been identified. However, from the points of view of the companies involved, the problems were solved and controls were instituted to prevent further outbreaks of these contaminations. These results were achieved by characterising the compounds involved, by GC retention times and sensory assessments, and then by the skilful use of the data the origins of the compounds were established. For the benefit of the food industry at large, it is still essential that these compounds be identified so that they may be added to the growing list of compounds known to cause off-flavours in foods.

When batches of chocolate confectionery were found to be contaminated with a catty-like off-flavour, in-plant studies indicated that some rolls of wrapping film were the immediate source of the problem. Analyses of extracts of the chocolate and film by GC combined with sniffing of the GC column outlet confirmed this finding and established retention times for the offending compound.

Subsequent analyses by GC–MS failed to identify it but on the basis of retention data the compound was not 4-mercapto-4-methylpentan-2-one, a compound well known to cause catty taints in meat and some packaged foods.[34] Continuing studies, using GC, showed that the compound was also present in only one type of varnish used in the printing of the wrapping films. This varnish differed from others in that it contained, as an additive, the titanium complex of pentane-2,4-dione. When hydrogen sulphide reacted with this additive an intense catty odour was produced. The reaction product did not have the same retention times as the off-flavour compounds, but was similar. Further in-plant investigations showed that a solvent used in the preparation of the varnish was occasionally contaminated with trace quantities of alkyl mercaptans. None of the original solvent remained and the supplier, although admitting that traces of mercaptans sometimes occurred in certain solvents, was unable to supply samples of the contaminated material. Studies involving the reaction of the pentane-2,4-dione complex with alkyl mercaptans are continuing; the aim is to identify the off-flavour compound. In this case it is important to reflect on the fact that the additive and the solvent, in isolation, would not have caused a problem, and that on many occasions the solvent would have been free of mercaptans. Unfortunately on this one occasion their combination resulted in the destruction of several days' production of a rather costly foodstuff.

In the second example in which the compound responsible for the off-flavour has not been identified, in-plant studies indicated that the source of an animal-musty off-flavour in bread was several batches of sodium stearoyl lactylate, used in the industry as an emulsifier.[35] Analyses of extracts of the bread and the emulsifier confirmed this finding and established retention times for the off-flavour compound. The problem was, however, complicated by the fact that two suppliers of the additive appeared to be producing products contaminated with the same off-flavour compound. The link between these companies was that both were using, as a raw material, stearic acid obtained from the same Asian supplier. Analyses of samples of stearic acid confirmed that the suspect material was also contaminated with the off-flavour compound.[34] In-plant examination of several Asian factories producing stearic acid from palm oil showed that only one was involved in the production of the off-flavoured product. In this plant some recycled residues were used and the source of the off-flavour compound was traced to these materials. Studies are continuing involving these recycled residues, with the aim of identifying the off-flavour compound. In retrospect the problem occurred because of the now appreciated unwise act of using recycled residues and the existing regulations used to define food grade chemicals of plant and animal origin.

With the problem involving the sporadic occurrence of a disinfectant off-flavour in canned beer, chlorophenol contamination was initially suspected. Analysis of the product showed that these compounds were not involved and that the cause of the problem was iodoform. In-plant investigations indicated that the problem only commenced with the introduction of an iodine-based sterilising agent and, since alkali was also used in the cleaning of the fermentation vessels and connecting lines, only a trace of ethanol was required to produce the off-flavour compound.[36] Subsequent laboratory studies showed that, on the addition of the sterilising agent to an alkaline solution containing ethanol, iodoform was precipitated. It was reasoned that, after sterilisation of the plant, small quantities of the above reagents could be either retained in the vessels or could be trapped in the connecting plumbing. This material, together with any iodoform so formed, could either be swept into the fermentation vessels, causing gross contamination, or carried along with the product to the canning line where only those cans filled first would contain enough of the compound to cause an off-flavour. The latter explanation could account for the sporadic nature of the problem. Significantly, the off-flavour problem disappeared from the processing line with the introduction of an iodine-free sterilising agent.

CONCLUSION

In this paper an attempt has been made to indicate the variety of compounds which currently are associated with off-flavours in foods, and to demonstrate the different origins and pathways by which these compounds can enter the food chain. However, in the future the publication of such information may be prevented. In most Western countries, governments are now requesting that publicly funded laboratories should in part finance some of their own activities, particularly in areas of applied research. The private sector will no doubt be willing to finance such studies, including those into the causes of food off-flavour problems, but in return it can be expected that this work will be carried out under a cloak of confidentiality. Large food companies and many private research associations already carry out investigations in this field but their findings rarely reach the food industry at large. This attitude has no doubt resulted in the reinvestigation of the same problem many times. It is a sobering thought to reflect on the cost to the community if research carried out in the 1970s at the TNO and FRI-Norwich laboratories, into the causes of mustiness in broiler chickens, had not been published. It is to be hoped that the food industry will quickly appreciate that, in areas affecting the quality of their products, they have everything to lose by remaining silent. Clearly the industry's cause and that of the consumer can best be served by the rapid dissemination of all pertinent data.

REFERENCES

1. WHITFIELD, F. B. (1983). *CSIRO Food Res. Q.*, **43**, 96–106.
2. WHITFIELD, F. B. and SHAW, K. J. (1985). In: *Progress in Flavour Research 1984*, Ed. Adda, J., Elsevier, Amsterdam, 221–38.
3. WHITFIELD, F. B. and TINDALE, C. T. (1984). *Food Technol. Austr.*, **36**, 204–14.
4. HURRELL, R. F. (1982). In: *Food Flavours*, Eds Morton, I. D. and MacLeod, A. J., Elsevier, Oxford, 399–437.
5. WHITFIELD, F. B. and LAST, J. H. (1986). In: *The Shelf Life of Foods and Beverages*, Ed. Charalambous, G., Elsevier, New York, 483–500.
6. ENGEL, C., DE GROOT, A. P. and WEURMAN, C. (1966). *Science*, **154**, 270–1.
7. CURTIS, R. F., LAND, D. G., GRIFFITHS, N. M., GEE, M., ROBINSON, D., PEEL, J. L., DENNIS, C. and GEE, J. M. (1972). *Nature*, **235**, 223–4.
8. GRIFFITHS, N. M. and LAND, D. G. (1973). *Chem. and Ind.*, 904.
9. NGUYEN, T. H. L., SHAW, K. J. and WHITFIELD, F. B. (1985). *Div. Food Res. Rep. Res. (Austr. CSIRO) 1984–5*, 25.
10. FREITER, E. R. (1979). In: *Kirk-Othmer Encyclopedia of Chemical Technology*, 3rd edn, Wiley, New York, 862–72.
11. WHITFIELD, F. B., NGUYEN, T. H. L., SHAW, K. J., LAST, J. H., TINDALE, C. R. and STANLEY, G. (1985). *Chem. and Ind.*, 661–3.
12. VOSS, R. H., WEARING, R. O., MORTIMER, T., KOVAES, T. and WONG, A. (1980). In: *Proc. 3rd Int. Congr. Industrial Waste Water and Wastes*, Stockholm, Sweden, 1–13.

13. WHITFIELD, F. B., TINDALE, C. R., SHAW, K. J. and STANLEY, G. (1984). *Chem. and Ind.*, 772–4.
14. JONES, C. R. and HART, H. V. (1959). *Res. Assoc. Brit. Flour Millers Bull.*, **9**, 61–2.
15. WHITFIELD, F. B., LAST, J. H., SHAW, K. J. and MUGFORD, D. C. (1984). *Chem. and Ind.*, 744–5.
16. GHOSH, B. L., PAL, P. N. and BASU, S. N. (1962). *J. Text. Inst. Trans.*, **53**, T577–8.
17. SALAM, M. A., ABDULLAH, A. B. M. and KHAN, N. N. (1977). *Bangladesh J. Jute Fibre Res.*, **11**, 57–64.
18. STANLEY, G. (1982). *Div. Food Res. Rep. Res. (Austr. CSIRO) 1981–2*, 25.
19. WHITFIELD, F. B., FREEMAN, D. J. and BANNISTER, P. A. (1981). *Chem. and Ind.*, 692–3.
20. FREEMAN, D. J. and WHITFIELD, F. B. (1983). *Div. Food Res. Rep. Res. (Austr. CSIRO) 1982–3*, 24.
21. WHITFIELD, F. B., TINDALE, C. R. and LAST, J. H. (1983). *Chem. and Ind.*, 316–17.
22. PIET, G. J., ZOETEMAN, B. C. J. and KRAAYEVELD, A. J. A. (1972). *Water Treat. Exam.*, **21**, 281–6.
23. WHITFIELD, F. B., FREEMAN, D. J., LAST, J. H. and BANNISTER, P. A. (1981). *Chem. and Ind.*, 158–9.
24. WHITFIELD, F. B., FREEMAN, D. J. and SHAW, K. J. (1983). *Chem. and Ind.*, 786–7.
25. WHITFIELD, F. B. and FREEMAN, D. J. (1983). *Water Sci. Technol.*, **15**, 85–95.
26. MILLER, A., III, SCANLAN, R. A., LEE, J. S. and LIBBY, L. M. (1973). *Appl. Microbiol.*, **26**, 18–21.
27. RADFORD, D. and DALSIS, D. E. (1982). *J. Agric. Food Chem.*, **30**, 600–2.
28. MCBRIDE, B. C. and WOLFE, R. S. (1971). *Biochemistry*, **10**, 4312–17.
29. FREEMAN, D. J., IZZARD, M. E. and WHITFIELD, F. B. (1985). *Austr. Fish.*, **44**, 35–6.
30. WHITFIELD, F. B., LAST, J. H. and TINDALE, C. R. (1982). *Chem. and Ind.*, 662–3.
31. ELSDEN, S. R., HILTON, M. G. and WALLER, J. M. (1976). *Arch. Microbiol.*, **107**, 283–8.
32. WHITFIELD, F. B., FREEMAN, D. J., LAST, J. H., BANNISTER, P. A. and KENNETT, B. H. (1982). *Austr. J. Chem.*, **35**, 373–83.
33. LAST, J. H., SHAW, K. J. and WHITFIELD, F. B. (1984). *Div. Food Res. Rep. Res. (Austr. CSIRO) 1983–4*, 26.
34. PATTERSON, R. L. S. (1969). *Chem. and Ind.*, 48–9.
35. LAST, J. H. and WHITFIELD, F. B. (1984). *Div. Food Res. Rep. Res. (Austr. CSIRO) 1983–4*, 26.
36. FIESER, L. F. and FIESER, M. (1956). *Organic Chemistry*, 3rd edn, Reinhold, New York, 156.

Index